W9-CFO-317

Global Broadcasting Systems

Robert L. Hilliard   Michael C. Keith

FLORIDA STATE
UNIVERSITY LIBRARIES

FFB 16 1996

Tallanassee, Florida

Focal Press
Boston Oxford Melbourne Singapore Toronto Munich New Delhi Tokyo

HE
7631
H54
1996

Focal Press is an imprint of Butterworth–Heinemann.

Copyright © 1996 by Butterworth–Heinemann

Ⓡ A member of Reed Elsevier group

All rights reserved.

No part of this publication may be reproduced, stored in a retrieval system, or transmitted in any form or by any means, electronic, mechanical, photo-copying, recording, or otherwise, without the prior written permission of the publisher.

∞ Recognizing the importance of preserving what has been written, Butterworth–Heinemann prints its books on acid-free paper whenever possible.

Library of Congress Cataloging-in-Publication Data
Hilliard, Robert L.,
   Global broadcasting systems / Robert L. Hilliard, Michael C. Keith.
       p.   cm.
   Includes index.
   ISBN 0-240-80197-0 (pbk.)
   1. Telecommunication.   2. Telecommunication policy.   I. Keith, Michael C.
II. Title.
HE7631.H54   1996
384—dc20   95-5690
                                                             CIP

British Library Cataloguing-in-Publication Data
A catalogue record for this book is available from the British Library.

The publisher offers discounts on bulk orders of this book.
For information write:

Manager of Special Sales
Butterworth–Heinemann
313 Washington Street
Newton, MA 02158–1626

10 9 8 7 6 5 4 3 2 1

Printed in the United States of America

Contents

Preface

Two major problems confront the researcher who wishes to write a book on the current global systems of telecommunications: (1) in most of the countries where the government tightly controls the media, candid and accurate information is not made available; (2) in most nations in the world, ranging from small third-world developing countries to large industrial countries like the United States, telecommunication systems are changing so rapidly that information that may be valid one day may be out of date the next. New stations, cable connections, and satellite transmission are adding to or replacing traditional terrestrial systems in many parts of the world. Countries that for decades operated government or public noncommercial systems are moving rapidly toward private commercial stations. Programming is changing rapidly from increasing international satellite coverage and program exchange throughout the world. Countries with developing television systems and limited resources for domestic production are importing foreign programs, mostly entertainment series from the United States, but also to a large extent from other countries with sophisticated video production such as Britain, France, Germany, and Spain. The proliferation of stations in many countries has resulted in competition for advertising support. Consequently, many countries that heretofore had programming of generally high cultural and artistic quality began importing programs that followed the U.S. commercial pattern of appealing to the lowest common denominator (LCD). In addition, real-time international news via satellite continues to affect the structure and operation of many systems. Much of what is new today is old tomorrow.

Further, we are moving faster and faster into a multimedia, narrowcasting age. With the convergence of technologies, all media increasingly relate to one another. Of the 7,000 or so languages in the world today, about 96 are usable for 96 percent of the population. What was once geographical and communications isolation for many groups has now moved to a global intercom. Some critics, in fact, are saying that with cyberspace communications, radio and television are virtually obsolete. In many countries well along the high-tech road, broadcasting as we have known it will likely soon be replaced by new multimedia interactive communications systems. But throughout most of the world, radio and television are still the principal means of mass communication for the present and for the foreseeable future. The plaint of a Bosnian refugee during the 1990s in former Yugoslavia puts it dramatically: "They came to our village at night. They took our televisions, our lambs, our calves. They took the best things we had!"

Mordecai Kirshenboim, Director-General of the Israel Broadcasting Authority advised us: "Be very fast in writing your book, because changes are happening so fast, whatever you are finding now will quickly be out of date."

Enrique Jara, Director of Media Products for Reuters Television, put it to us this way: "Is this the moment in history when the building blocks of slow historical change have reached a new plateau and new usable roads are ready? Is this the moment of history when previous ways of doing things are no longer applicable?"

We don't know. Perhaps one can never know, and all that can be done is what we are attempting to do in this book: record this moment in history, fallible and dated as it may be, to the best of our ability. If what we have found and put down here has changed by the time it is being read—or even by the time it has left our computer screen for the printed page—we have at least, we hope, provided an overview that offers some perspective and maybe a guideline or two for continued progress.

We wish to acknowledge the staff of Butterworth-Heinemann/Focal Press, through whose offices this book has passed on its way to publication. We appreciate the special contributions within the text of Lynn Christian, Indra de Silva, Marilyn Matelski, Sejid Qaisrani, Bill Siemering, and Nancy Street. Our appreciation goes too to our colleagues and students who peppered us with questions, ideas, and materials as we taught areas of international communications. And our special thanks to Carla Brooks Johnston, who permitted us to interview her and to use material from her interviews with many of the key people referred to in this book, as she researched her own new book, *Winning the Global TV News Game.*

Robert L. Hilliard and *Michael C. Keith*

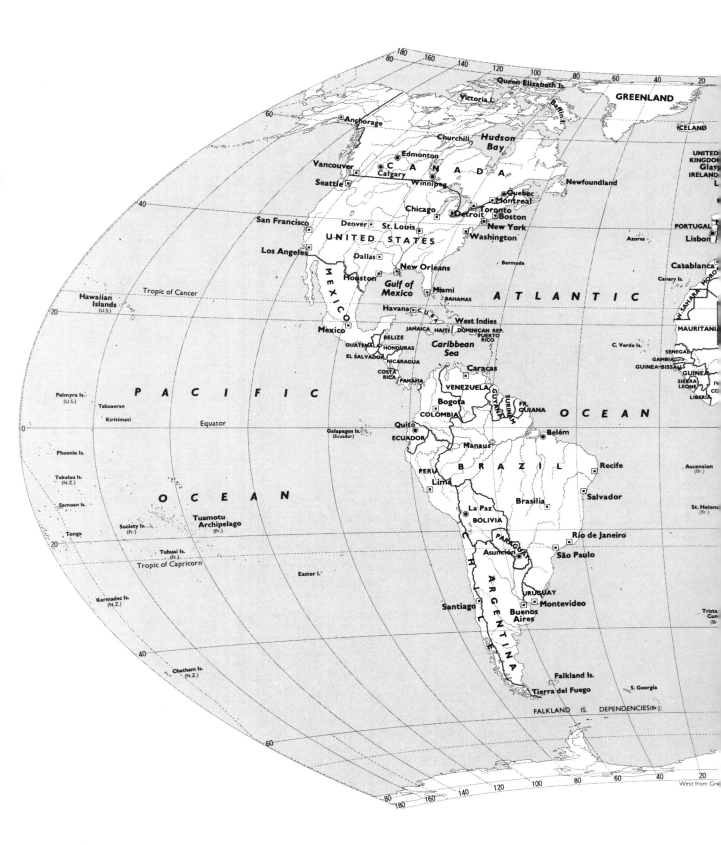

Projection: Hammer Equal Area

A R C T I C   O C E A N

R 20 40 60 80 100 120 N 140 160 180

New Siberian Is.

Novaya Zemlya

SWEDEN FINLAND
Helsinki
St. Petersburg (Leningrad)
Stockholm
Copenhagen Moscow
POLAND Warsaw
Berlin UKRAINE Kiev
Vienna
ROMANIA
BULGARIA Bucharest
GREECE Istanbul
TURKEY

Arkhangelsk

RUSSIA

Yekaterinburg   Novosibirsk

Irkutsk

KAZAKHSTAN

Ulan Bator
MONGOLIA

Bering Sea

Vladivostok   Sapporo
N.

Baku
UZBEKISTAN Tashkent
TURKMENISTAN KIRGHIZIA
TAJIKISTAN

Tehran
AFGHANISTAN
Kabul
Islamabad
Lahore

C H I N A

Peking
KOREA JAPAN
Dalian S. Tokyo
Pusan Osaka

PACIFIC

Athens
Mediterranean Sea
Tripoli
Alexandria   Cairo

SYRIA Baghdad
ISRAEL IRAQ
JORDAN
KUWAIT

IRAN

NEPAL
PAKISTAN
Karachi   Delhi
BANGLA-
DESH Dacca

Chungking

Shanghai

Hong Kong

Tropic of Cancer

Wake I.
(U.S.)

20

LIBYA   EGYPT

SAUDI
ARABIA   OMAN
Mecca

BAHRAIN
U.A.E.

Ahmadabad

I N D I A
BURMA
(MYANMAR)

Calcutta
Hanoi

TAIWAN

Northern
Marianas

O C E A N

NIGER   CHAD
ney Kano
NIGERIA Ndjamona
gos
CAMEROON CENTRAL
AFRICAN
REPUBLIC
Douala
RIAL
EA GABON

Khartoum
SUDAN

YEMEN

Arabian
Sea

Bombay

Madras

Bay of
Bengal

Rangoon

THAILAND
Bangkok
Phnom
Penh

VIETNAM
CAM-
BODIA

Manila

PHILIPPINES

Ho Chi Minh
City

Marshall Is.

Addis Ababa
DJIB.
SOMALI
ETHIOPIA

Colombo   SRI LANKA
Maldives

MALAYSIA BRUNEI
Kuala
Lumpur
Singapore Borneo

Federated States
of Micronesia

Kiribati
Nauru

Equator   0

ZAIRE
UGANDA
KENYA
RWANDA Nairobi
BURUNDI
CABINDA
Kinshasa
TANZANIA Mombasa
Dar-es-Salaam

Mogadishu

Seychelles

I N D I A N

Padang

Sumatra

INDONESIA
Jakarta   Surabaya

New PAPUA
Guinea NEW GUINEA

Solomon Is.

Port Moresby

Tuvalu
(Ellice Is.)

Luanda
ANGOLA
ZAMBIA
Harare
NAMIBIA ZIMBABWE
BOTSWANA
Johannesburg SWAZ.
SOUTH LES.
AFRICA Durban
Cape Town

MOZAMBIQUE

O C E A N

Antananarivo
MADAGASCAR
Mauritius

Darwin

Alice Springs
A U S T R A L I A

Coral
Sea

Vanuatu

Fiji

New
Caledonia
(Fr.)

20

Tropic of Capricorn

Crozet Is.
(Fr.)

Kerguelen Is.
(Fr.)

Perth

Adelaide
Melbourne
Tasmania

Sydney
Canberra

Hobart

NEW
ZEALAND

Auckland

Christchurch

40

Dunedin

am Greenwich

S O U T H E R N   O C E A N

20   40   60   80   100   120   140   160   180

60

1:88 000 000

Copyright, George Philip & Son, Ltd.

The World Telecommunications Revolution

"We are witnessing the revolution of the empowerment of the media consumer," Reuters Television director Enrique Jara told us in early 1994. Communications technology is making everything directly available to the consumer, eliminating the traditional media editors, Jara says. Consumers can select from a total menu of a multimedia TV screen source. Jara states that the consumer, not the editors as in the past, will choose the distribution channel; the job of the editor will be to guarantee the integrity of the material to be stored.

While the empowerment of the consumer may not yet be a widely supported goal, and is certainly not the aim of the media distributors, many leaders in the field of international telecommunications believe that the future is a multimedia retrieval system for everyone. Broadcasting as we know it will go through a transition from the current limited-area terrestrial systems to multichannel, worldwide, satellite-to-cable retrieval systems. While agreeing with the dramatic changes occurring in global systems, some media executives, such as Ian Frykberg of British Sky Broadcasting, Ltd., operator of the international SKY channel, believe that the combination of satellite and cable distribution won't entirely replace traditional terrestrial systems because the great diversity in populations and cultures, even in the fast-approaching "global village," will require special-interest, narrow-need programming.

Jara is one of those who emphasizes the importance of diversity. "There is more diversity in the world than there is a global village," he says. The emerging nationalism in countries throughout the world in the 1990s, as evidenced in Russia and other states of the former Soviet Union, suggests that one cannot develop programs suitable for everyone in every country. While Jara believes terrestrial distribution will be replaced by a combination of satellite and cable, he also believes that individual countries will have to maintain or develop their own media around their own individual cultures.

A somewhat different view of the retention of individual cultures is taken by Dr. Ziad Rifai, a university professor and communications officer with UNICEF. He cites the influence of one international news channel, CNN, as an example of how the look, content, and methods of reporting news are becoming similar in many countries throughout the world through the copying of CNN approaches. He believes that this is reflected in other types of programming as well. American culture is a powerful influence in other cultures through the introduction of its TV programming into other countries. Many countries are losing their cultural identities to the "New TV Culture." There seems to be one emerging international culture, Rifai told us, because international satellite programming has an impact on both regional and local (individual country) satellite programming.

A similar view of a global communications village, but with an emphasis on the growth and strengthening of local stations, is held by Nachman Shai, Director-General of Israel's Second Television and Radio Authority, which went on the air in November of 1993. Shai noted to us the "fly-way system" of broadcasting live directly to satellite, which grew during the Gulf War. He believes that in the future all wars and international news will be covered in this way, bypassing censor intervention, thus resulting in a global news village. Shai says, "This will allow local stations everywhere to carry international news. No network will have an advantage anymore." While this kind of coverage is technically possible now, political factors continue to impede its implementation.

In another part of Jerusalem, Mordecai Kirshenboim, Director-General of the Israel Broadcasting Authority (channel 1) takes a somewhat different view. While satellite and cable will have a profound effect on world communications, he told us, he believes that the entire world will eventually have about 500 channels with a heavy emphasis on interactive TV and pay-per-view, and no channel will likely have ratings higher than 2 or 3.

Robert E. Burke, president of World Television News, headquartered in London, agrees with the coming growth in the number of local stations. "With more and more countries developing systems free of government control," Burke told us, "we are seeing more and more stations throughout the world." Globally, the growth of technology is drastically changing media systems, although cost factors have made such changes greatly uneven. New distribution systems are restructuring U.S. broadcast TV; Burke believes that Europe will be going through the same process in about five years.

Johan Ramsland, editor of BBC World Service Television News, points to the growth of international channels. As an example, he noted to us that Rupert Murdoch already has the STAR and SKY satellite channels serving much of the world (Europe, Asia, and Africa), and that he owns the FOX television network in the United States and part of Australian network channel 7. "Murdoch already has parts of a world jigsaw," Ramsland says, "on the way to creating an international global service." This is confirmed by Ian Frykberg, head of news and sports for Murdoch's SKY channel. "Murdoch wants SKY NEWS in every continent in the world," Frykberg told us, "and we're working toward it . . . toward a separate international channel."

However, Ramsland notes, too, that the end of the cold war and the bipolar world is resulting in greater nationalism in individual countries, and he wonders how many global networks can ultimately survive.

LAUNCH OF GLOBAL COMMUNICATIONS

In the early 1960s, when communications satellites became possible, practitioners and policy-makers were already promoting the concept of worldwide communications systems. In 1961 a news release from RCA quoted from a speech by the Chairman of RCA, David Sarnoff:

Ten years hence—if vigorous foreign growth continues—there will be TV stations in virtually every nation on earth telecasting to some two hundred million receivers. An audience of a billion people might then be watching the same program at the same time, with simultaneous translation techniques making it understandable to all. In a world where nearly half of the population is illiterate, no other means of mass communication could equal television's reach and impact on the human mind.

That same year, in the *Christian Science Monitor*, Neal Stanford wrote: "What the industrial revolution did to the society and economy of our forebears is nothing compared to what the space revolution we are now experiencing is going to do to our economy and society in the next few decades."

Since the early 1970s many conferences and organizations, such as the International Association for Mass Communication Research, have stressed again and again how developing technology has required increasing cooperation among national systems, and how such increasing cooperation has made it possible for the development and practical application of new global communications technology.

Through the 1970s and 1980s numerous individuals, and formal and informal associations, in many parts of the world, pushed for implementation of global television. Many had limited agendas, but almost all saw positive political, social, and/or educational goals that would bring the peoples of the world closer together through mutual understanding, sharing, and solutions to common needs and desires. But most were ahead of their time; for example, the Committee for International Tele-Education (CITE). Headquartered in Washington, D.C. with international membership, CITE held conferences and published books and papers, but was unable to attract the monetary support necessary to continue its existence into the 21st century, when technology, economics, and politics will, indeed, come together to prompt mutual learning and exchange among students throughout the world.

In general, telecommunications practitioners agree that there is a new world market developing, in which a greater demand for programming will create new alliances—"partners," as most put it—internationally and regionally, to produce and provide programming that one producer/distributor or one station or system alone cannot do as effectively or economically.

The new world order of programming will both emanate from and contribute to the development of regional and international systems, which in turn will affect the present systems of individual countries. There are drastic changes now occurring in the way in which individual countries' telecommunications

systems have been operating for the past seventy years since radio made its way onto the world scene. Larger regional and international conglomerates are moving into well-financed multimedia operations that include video, audio, and print.

POLITICAL AND SOCIAL IMPLICATIONS OF WORLD TELECOMMUNICATIONS

The political and social implications of telecommunications, in relation to reflecting and affecting democratic movements, are significant in terms of changing systems and system changes. Robert E. Burke, President of Worldwide Television News, told us he believes that a significant by-product of technological communications change "is that TV and other media, including the FAX machine, are great for democracy. They make it more difficult for governments to keep out information they don't want people to have." One of the many conferences in recent years along these lines is an international meeting developed in 1994 by Professor Dov Shinar, Dean of the New School of Media Studies of Hebrew University in Tel Aviv: "Communications as Instruments of Peace." As early as 1970, the Institute for International Communication, then the International Broadcast Institute, held a conference on "The New Communication Technology and its Social Implications." The discussion included the potential impact of "the impending communication revolution which may, within a decade, completely change the established pattern of broadcasting." The conference explored the key issues that are just now, in the 1990s, demanding resolution, including private vs. public control, the phasing out of existing technology, freedom of speech and the free flow of information, multichannel development and diversity of services, information overload, the needs of developing countries, and the impact of international communications exchange on traditional cultures.

In 1969 a graduate-degree-granting institution was established in Washington, D.C. with the appropriate name of The International University of Communications (IUC). Enrolling students from various countries of the world, the IUC operated on the premise that technology had made the global communications village not only possible, but inescapable, and that the public had to learn how to control the media for its own social good, rather than let the media control the public. The IUC's students' major projects were oriented toward demonstrating the use of communications to affect public policy on national and international levels, which would concomitantly require changes in traditional systems of production, distribution, and reception.

Nearly a quarter of a century later, in 1993, the International Academy of Broadcasting (IAB) was founded in Switzerland with a somewhat similar agenda, as its statement of purpose reveals:

> The world of broadcasting is becoming more and more competitive year by year. Broadcasting is also becoming more international, requiring a knowledge and an understanding of structures throughout the world. The arts and sciences of broadcasting represent today a complex interweaving of different technological, creative, cultural, commercial and aesthetic disciplines. The leaders in this environment have to combine a very high degree of expertise in one or more particular fields, whilst maintaining a broad and comprehensive knowledge of the whole domain of broadcasting. With the changing map of the world broadcasting scene and with all the technological breakthroughs, broadcasting in the next century will differ considerably from what we know now; consequently, it will require a new breed of leaders. The aim of the International Academy of Broadcasting is to prepare those leaders to assume this herculean task.

The IAB is supported by a host of international broadcast institutions, among them the European Broadcasting Union (EBU), the International Television Symposium, the Asia-Pacific Broadcasting Union (ABU), the Organizacion de la Television Iberoamericana (OTI), the International Telecommunication Union (ITU), and UNESCO.

Newton Minow, who as Chair of the Federal Communications Commission so strongly forwarded the Kennedy-era public interest reorientation of government, warned broadcasters in a 1962 speech that the coming of worldwide television broadcasting required mutual responsibility for content among the nations of the world. "Your government will not and cannot monitor or censor your world programs . . . that's going to be the job of your conscience and your character. The penalty for irresponsibility will be more serious for the nation than the revocation of a station license."

Writing in the spring 1993 issue of the *Journal of Broadcasting & Electronic Media* on early policy concepts regarding global television, Michael Curtin stated that "global television, like the newspaper of the 19th century, implied a community of address and a clocked consumption of information and images. Yet unlike the newspaper, television promised to bring people together across boundaries of the modern nation-state on a regular basis."

ROADBLOCKS TO INTERNATIONAL ELECTRONIC MEDIA

The question of censorship is a vital one, revolving around each country's perception of whose ox is being gored. Unless international agreement can be reached concerning censorship, free international telecommunications is not possible. In 1990, the Stanley Foundation held a conference on strategies for peace that included a session on international communications and censorship. The premise of the discussion was that the new technologies were making it easier for people everywhere to communicate with each other, individually as well as on a mass basis, and at the same time were making it more difficult for governments to control the content of such information exchange. The conference report, "Regulating International Communications Without Censorship," issued at the Thirty-first Strategy for Peace U.S. Foreign Policy Conference, October 25-27, 1990, stated: "Freely mobile communications can educate, re-create, and inform policy; communications regulated with narrow nationalistic rather than universal objectives can provide a new, intolerable form of censorship." The conference report noted also that, historically, U.S. foreign policy has not clearly recognized the importance of international telecommunications to the progress of global educational, economic, and social goals. Clearly, this shortsightedness is shared by almost all other countries in the world, creating a global roadblock for effective global telecommunications. The report noted the universal positive and negative impacts of instant, unfettered information flow in many critical situations, citing as a negative example how a report of cold weather in Brazil could raise coffee prices instantaneously throughout the world, and as a positive example how a report of an earthquake could result in immediate aid from all parts of the world to the victims. While stating the value of international exchange of programs as a step toward cultural diversity and understanding in all countries, the report expressed concern about the potential control of satellites by sending and receiving nations for the purpose of cultural imperialism. It noted the need for an infusion of funds to provide equipment to have-not countries, particularly to people in remote areas who are not served by telecommunications companies because of the cost of providing such services. Richard Stanley, president of the Stanley Foundation, suggested one solution to the problems "presented by integrating global communications systems . . . laws and institutions must be reformulated to encourage the development of services and access to information."

ADDRESSING THE ISSUES

Several key issues emerge as global telecommunications systems develop. One is the necessity of being certain that all countries and persons desiring any available global communications are able to receive them. This implies the need for political freedom and technical resources to receive and send information, and a pricing structure suitable to the economy of any given receiving country or sub-area. Global telecommunications must reach the maximum number of potential users at an affordable price. A second issue is the need for global cooperation to maximize the use of available technology; this includes agreements on international standards of compatibility, and an integrative, optimally efficient use of all available media, from satellite to cable to terrestrial broadcast to microwave to fiber optics to the technologically possible, but not yet economically feasible, three-dimensional laser communications. Third, and dependent on achievement of the first two goals, all countries, developed and developing, first and third world alike, will have to be treated equally in regard to global telecommunications participation.

Yet the need for regional and individual protection against what may be perceived as economic and cultural imperialism is clear. For example, in the development of free trade among its twelve member countries in the early 1990s, the European Economic Community included a directive protective of its broadcasting activities. Then referred to as the EC, the alliance has since become the European Union and now is referred to as the EU. Principally, the EU required its members to devote a majority of their broadcast time to programs of European origin, to limit the total amount and placement of advertising, to restrict advertising of products such as tobacco and alcohol that are deemed harmful to the public, and to preclude programming that might be harmful to minors (including pornography and gratuitous violence) or that might be considered racist or sexist in nature. While such requirements can be justified on one hand by countries seeking to protect their economy, culture, and moral and ethical principles, these restrictions may be challenged on the other hand by other countries as inhibiting free information exchange and free speech. The United States, for example, through its Trade Representative, Carla Hills, called the EU broadcasting directive "blatantly protectionist and unjustifiable." Jack Valenti, president of the Motion Picture Association of America, asked, "Is the real game not culture, but commerce?"

If the western countries, with their tradition of free information exchange and cultural cooperation, can be thus split on the basis of economic interests, will other regions' and countries' diverse cultures and economic histories make it impossible to establish the kind of cooperation necessary for effective global telecommunications systems?

A case in point, concerning the fear of what foreign television might do to an individual country's culture, is epitomized in a comment by Israel Foreign Minister Shimon Peres in late 1993: "The greatest threat isn't a military invasion, but a cultural invasion, and cable television is more dangerous to our identity than the intifada. Television knows no boundaries and for some youths the most important woman isn't Rebecca or Sarah [biblical figures], but Madonna." On the other hand, the full freedom of broadcast communications has played a significant role in conveying messages of peace to many countries and promoting cooperation and conciliation instead of war. Under a headline entitled "handshake for history," the *Boston Globe* of September 14, 1993, asked, "Has the video camera captured anything as dramatic as Yasser Arafat's handshake with Yitzhak Rabin?" The *Globe* stated that only television can capture such "a moment of pure transcendence, where differences seemed to melt away, be they differences between Arabs and Israelis or Democrats and Republicans." Freedom of information made it possible in June, 1991, for an independent, quasi-underground radio station in Russia, Moscow Echo, to report the events in Lithuania as they happened, information that served as a catalyst for political change throughout the entire Soviet Union.

A BENEVOLENT FORCE

Communications already are being used internationally for common humanitarian goals, aside from political or economic considerations. For example, The International Federation of Red Cross and Red Crescent Societies has determined that global communications systems can play important roles in disaster relief and assisting victims of natural and human-caused disasters. In 1991, representatives from twenty-five countries issued "The Tampere Declaration on Disaster Communications," in which they stressed the necessity of international cooperation to enhance national and international communications systems to reduce the loss of life and damage to property resulting from disasters. In a 1992 report issued by the Northwestern University's Annenberg Washington Program for Communication Policy Studies, entitled "Disaster Communications and Information Management," Dale Hatfield stated, "Appropriate use of telecommunications and information systems . . . is apt to result in more timely, efficient and effective delivery of scarce humanitarian aid, thus saving lives and property."

In late 1993, the World Health Organization emphasized the need to use communications to provide health education to the public. It called for the cooperation of media and government to use drama and other program formats to improve health practices and conditions internationally.

A further effect of worldwide telecommunications systems is the concomitant development of worldwide personal communications. While this book does not deal with personal communications services, it is important to note that such services are, to a great extent, incorporated with the technology used for mass media services: satellite, cable, fiber optics, and microwave (wireless cable). At the present time, digital radio-based technology transmits personal voice and data services over broad areas. Eventually, any person anywhere on the globe will be able to communicate instantly and personally, point-to-point, with any other person on the globe, any place, any time. In 1995 global interpersonal communication was well on its way with the increasing use of Internet by those in countries throughout the world who had the monetary means for computers and services to do so.

THE NEW TECHNOLOGIES

While political and economic issues are the most critical in international telecommunications in the long run and are the keys to global cooperation and coordination in the use of media, the initiation of movement toward the global communications village is made possible by changing technologies. Distribution systems are the bases for outreach. Satellites are the means for expansion (Figure 1-1).

Only a decade or so ago, even the countries with the most sophisticated

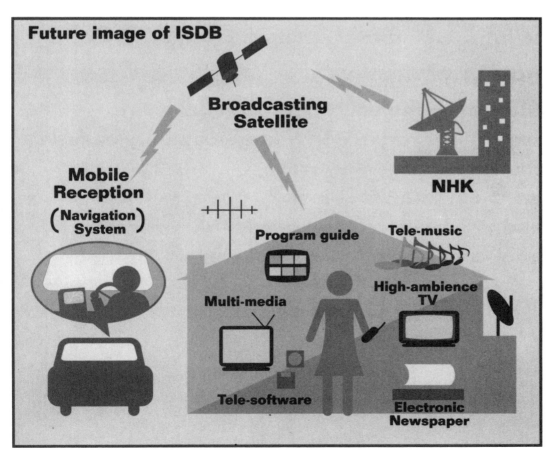

FIGURE 1-1
The new telecommunications technologies improve the quality and service of broadcast companies.
Courtesy NHK.

communications technologies were relying largely on terrestrial distribution. Over-the-air relays extended signals from antenna to antenna. Today satellites are used in an increasing number of nations not only for international exchange but for domestic distribution (Figure 1-2). Microwave dissemination has expanded point-to-point communications. Cable has provided for not only dozens but hundreds of signals to be received simultaneously at a given point. Fiber optics have enhanced not only capacity, but the quality of signals. Direct Broadcast Satellite (DBS), although growing more slowly than anticipated, is nevertheless moving at a fast enough rate to warrant the establishment of more and more regional satellite distribution systems. High-definition television (HDTV) can make every TV screen into a movie theater (such are manufacturers' claims) and further enhance the importance of satellite delivery. Digital broadcasting promises, by the beginning of the 21st century, to create a level playing field for all radio distribution systems (although there are skeptics, mainly among AM broadcasters) and guarantee the highest quality of audio reception for every listener.

INTERNATIONAL CONFERENCES

With the expanded use of telecommunications satellites around the globe and a host of new and evolving technologies, broadcasters are gathering in Europe to share ideas and promote their nations' electronic wares. The global privatization of electronic media has created a bull market for manufacturers of communications equipment and commercial broadcast management expertise and consultation from the United States and elsewhere.

In 1992, the first Montreux International Radio Symposium was held, and in attendance were delegates from

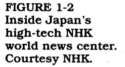

FIGURE 1-2
Inside Japan's
high-tech NHK
world news center.
Courtesy NHK.

dozens of nations in Europe and from other parts of the world, who possessed a vested interest in their individual country's developing and expanding broadcast systems (Figure 1-3).

The primary purpose of the Montreux gathering, says European Broadcasting Union (EBU) president Jean-Bernard Munch, was "to share professional knowledge and the sense of community of all those working in the broadcast media, especially now as the European media landscape is in the process of rapid change, driven by deregulation, technological advances, and market forces."

A member of the Montreux International Symposium committee, Lynn Christian, told us that he expects the conference and exhibition to grow dramatically in the years ahead, due to the rapid expansion of private commercial broadcast stations throughout the world. Says Christian, "A great need exists for technical and management information and resources as nations engage in a free enterprise system of radio and television."

Thousands of broadcasters from around the world attend the symposia in Montreux, and officials note that the number of participants rises sharply with each conference every two years. While the entire program is presented in English, the majority (80 percent) of those attending are from non-English-speaking countries.

Among the most highly attended panels at Montreux in 1994 were "The Co-Existence of Public and Private Radio,"

FIGURE 1-3
Broadcasters from around the world attend the conference in Montreux. Courtesy Lynn Christian.

"Central European Radio After Communism," "The Future of International Programme Exchange," "Is the Format Model Still Valid in Europe?" and "The Future of International Radio Service." More than three dozen symposiums and workshops were offered. As of this writing, gatherings of a similar nature are being planned in Asia and Africa.

Meanwhile, broadcasters from dozens of countries annually attend the U.S. conferences held by the National Association of Broadcasters (NAB) and the Radio Advertising Bureau (RAB). In a June 1, 1994 article in *Radio World,* RAB president Gary Fries observed that "the sudden climb in foreign attendees signals a new sense of shared challenges, shared opportunities and shared learning opportunities in the global broadcasting community." Fries goes on to note that "the recent birth of commercial radio in so many countries has put an increasing number of broadcasters in the same boat when it comes to finding ways to market their stations, audiences and marketing expertise to advertisers."

U.S. broadcast entrepreneurs have headed to Europe, Asia, and Africa in droves to pursue one of privatization's foremost goals, the generation of profit through advertiser-supported programming.

2 World Systems Overview

While there are literally hundreds of millions of television and radio sets in use throughout the world, their distribution is markedly uneven. United Nations' figures showing sets in use in the five major regions in the world illustrate the differences between current have and have-not areas and nations.

Throughout the world, there were about 375 radios and 150 TV sets per thousand people in use at the beginning of the 1990s. In North America, however, there were over 2000 radios and 800 TV sets per thousand people. In Africa, the figures were about 170 radios and 30 TV sets per thousand. South America had some 340 radios and 150 TV sets per thousand. In Asia, the figures were about 185 radios and 55 TV sets per thousand. In Europe, about 690 radios and 375 TV sets were in use per thousand people. And in the Pacific Basin, there were some 990 radios and 415 TV sets per thousand. Clearly these disparities must be corrected before a truly global telecommunications system can achieve the goals of equal access and unimpeded interchange for all people.

A further problem is that ownership figures, even when accurate, may be misleading. In many developing countries radios may be available, but not the batteries needed to operate them. International communications consultant Bert Cowlan points out that in many nations teachers are provided with radios for use in their classrooms. However, the price of a battery, which may be equivalent to 50 cents U.S., is prohibitive in many economies. A reduction in excise taxes on battery imports, Cowlan suggests, would go far to make radios operable in many third-world countries.

NORTH AMERICA

United States

Telecommunications systems in North America are plentiful and sophisticated. The United States dominates communications on the continent as, indeed, it does on the entire globe. While it did not have the first broadcasting system in the world, the United States has, throughout the century, been the leader in technical innovation and world program distribution. (The Netherlands had the first regularly scheduled radio station on the air in 1919, and the United States followed in 1920; the United Kingdom had the first successful public test of television in the world in 1926, and the United States followed in 1927.)

In the United States, by 1990 virtually 100 percent of all households had radios and 99 percent had television sets. In 1993 there was an average of one TV set per 1.3 persons and two radios per persons. Cable had reached over 60 percent of the nation's TV households, and the majority of the public had VCRs. The United States has more individual stations on the air than any other country, by virtue of its system of essentially private ownership of the broadcast media. Unlike most governments, the government does not own or operate any domestic stations that broadcast to the public. A system of noncommercial, or public, radio and television stations includes ownership by some state and local governments. All of these stations are licensed by the Federal Communications Commission, authorized under the Communications Act of 1934—amended many times since to include emerging technologies. The FCC regulatory and licensing authority is a result of the so-called "scarcity" theory, which essentially notes the limited number of frequencies available for telecommunications and requires a government agency to preclude monopoly operation of the media and to guarantee that licensees operate in the "public interest, convenience, and necessity."

The commercial system is totally dependent on advertising, and is responsible to the public only to the extent that it may voluntarily wish to consider the public interest on the basis of something other than the ratings of its programs—which it rarely does—or to the degree that the Federal Communications Commission, in any given political administration, decides to exercise a strong regulatory function in the public interest.

Other telecommunications distribution systems in the United States include microwave (sometimes called wireless cable), and Direct Broadcast Satellite (DBS). While the so-called "electronic information superhighway," with literally hundreds of interactive channels, is now being developed, the principal distribution system in the mid-1990s still remains individual station terrestrial dissemination, provided by broadcasting and cable networks.

In 1963 the United States established COMSAT (Communication Satellite Corporation) to coordinate space transmission and, as a result, quickly became the largest satellite distributor of TV programs worldwide. Commercial programmers buy satellite capacity to distribute U.S. programs to other countries' systems and stations and to

individually owned satellite receiving antennas throughout the world. These programs are principally entertainment shows, such as sitcoms and action-adventure drama series. In the early 1990s a news service based in the United States, CNN, became the dominant worldwide news channel, prompting many other news operations throughout the globe to copy CNN's style and approach. The U.S. domination of worldwide programming—albeit desired by the receiving countries and their viewing public—has created cries of cultural imperialism.

As do many other countries, the United States also reaches out through its overseas government broadcasting. The Voice of America, established in 1942, beams shortwave radio broadcasts internationally. Three U.S. radio services of the "cold war" period, Radio Liberty and Radio Free Europe (broadcasting to eastern European countries) and Radio Marti (broadcasting to Cuba) had outlived their usefulness and by 1995 had been either phased out or had their operations considerably reduced.

U.S. influence on the rest of North America, especially its neighbors to the north and south, Canada and Mexico, is great. While it does not control telecommunications in either of those countries, its policies and practices have a pronounced impact. Periodic border agreements are necessary, principally to guarantee the integrity of the frequencies assigned to the respective countries, but also to control the competitive advertising that appears on stations whose signals obviously can be seen and heard across the borders and that frequently lure customers across the borders to the detriment of the merchants in their own countries.

Still considered the focal point of communications production and distribution in the world, the United States has moved from limited-range terrestrial signal distribution to microwave relays to cable expansion to domestic satellite distribution to both stations and cable systems to a slow but gradual DBS industry. The early community antenna television systems (CATV) usually had a limit of twelve channels. These were sufficient then, inasmuch as the function of coaxial cable distribution to homes in a given area was to use a master antenna to provide a clear picture of over-the-air TV channels, channels that otherwise could not be received clearly. As cable technology, including fiber optics use, provided greater channel capacity, the relatively few broadcast stations that could be picked up in any given area were joined by cable networks, operating in the same way as broadcast networks. They distribute their signals by satellite directly to the local cable system, just as broadcast networks send their signals to their local station affiliates. The United States' five national TV broadcast networks, American Broadcasting Company (ABC), National Broadcasting Company (NBC), Columbia Broadcasting System (CBS), Public Broadcasting Service (PBS), and Fox, were joined in early 1995 by two fledgling networks, Warner Brothers Network (WB) and United Paramount Network (UPN). These are joined by literally hundreds of cable networks, most of them specialized.

Music, shopping, weather, gardening, religious, public affairs, news, education, feature films, and specialized kinds of formats are the sole or principal orientation of many cable networks, and many others concentrate on a particular genre, such as adventure entertainment, sitcoms, cartoons, or women's issues. Until the information highway of hundreds of channels becomes a reality as anticipated in the late 1990s, there are considerably more cable networks than channels available on any given cable system.

High-power satellites made DBS services possible; these are expected to grow as the cost and size of receiving dishes are drastically reduced. By 1994, a satellite dish the size of a dinner plate could receive signals in many cases just by being attached outside a residence window. As stated in the preceding chapter, in 1993 the FCC authorized the development of compatible high-definition television (HDTV). However, the price of HDTV sets is likely to be relatively high, and the system's compatibility with current receivers under the old National Television System Committee (NTSC) standards suggests that it could be an additional two decades before most of the country is receiving high-definition transmissions. For domestic purposes, some critics believe that fiber optics will compete successfully with direct satellite delivery. Traditional cable copper wire is being replaced in many parts of the country by the glass filaments of fiber optics, which provide superior signals, greatly enlarged signal capacity (estimated at 250 times that of copper wire), and lower cost. Meanwhile, cable researchers have developed a means to expand existing channel capacity through digital compression, a process that yields multiples of existing signal paths.

Canada

While from a world view the United States and Canada are in essentially the same geographical region and have many common interests and practices, their systems of telecommunications are somewhat different, as are, to an extent, the programming needs of their populations. However, their systems of distribution have gone through similar developmental processes. The set ratio in Canada is one TV per 1.7 persons and one radio per 1.2 persons.

Canada's population (under 30 million, about one-eighth that of the United States) is principally concentrated along the U.S. border. Its remote northern areas make it difficult to reach the entire population by television unless the potential viewer has electricity and a satellite receiving antenna. Radio covers the entire country. Another problem is the terrain, which makes it difficult for over-the-air broadcasting to reach everyone. Canada has solved that problem with one of the most extensive cable systems in the world, reaching more than 70 percent of its population. Canada's communications are controlled by the Canada Radio and Television Board, and their dissemination is through the Canadian Broadcasting Corporation (CBC), an independent, state-financed organization similar to the British Broadcasting Corporation (BBC). Because of the ethnic composition in the country, broadcasting is done in two languages: English and French. In addition to the CBC, there are private networks and local and regional stations, which operate like those in the United States. While the CBC tries to maintain programs of high cultural quality, it faces a worsening battle against popular entertainment shows from the United States. More than 80 percent of the English-speaking population is wired into private cable networks, which principally carry U.S. sitcoms and other light entertainment. Like many countries in other parts of the world, Canada is concerned about how the influx of U.S. programs is affecting its national culture and identity. In the early 1990s it was considering protectionist measures, similar to those adopted by the EU, and ways to encourage more Canadian production that will have popular appeal.

Before satellite and cable combined to achieve national coverage, Canada used microwave relays to carry the terrestrial signals across the country. However, the difficulty of constructing relay stations in some areas, and the limitations of microwave signals (due to their limited range and the effects of weather) made the microwave relay solution less than satisfactory. Canada's first use of cable was similar to that of the United States— a master antenna system designed to bring weak signals into an area's homes by coaxial cable. As cable technology has advanced and satellite delivery to cable systems have become standard, Canada has developed an extensive cable system. Canada's size and its rural character of isolated villages and towns made satellite technology not only beneficial, but necessary, for effective distribution.

Central America

North America includes Central America, although the Spanish-speaking lower part of the continent is frequently arbitrarily disassociated from the principally English-speaking upper part. As it does in many other areas of endeavor, the United States has strong influence and, to an extent, economic domination over telecommunications in Central America. The developing countries south of the U.S. border are strongly dependent upon foreign capital and technology in their telecommunications. The U.S. telecommunications industry, as well as other industries, in particular the three major broadcasting networks, NBC, CBS, and ABC, have been involved for some time in Central American broadcasting. Stations in most of these countries are heavily dependent upon U.S. programming and on advertising from U.S. multinational companies. Many of these countries are worried about the "cultural imperialism" represented in the importation of U.S. programming. The poor economic conditions of what are essentially third-world countries make it difficult, if not impossible, for many of the Latin American countries to produce their own programming. Every Central American country has at least one earth station to receive foreign programs via satellite.

Telecommunications development in most of the other North American countries, however, is dependent not only on a country's size and economic condition, but to a great extent on its political orientation to the United States. Those countries that have been in the good graces of the United States and have cooperated with U.S. business interests in the development of their investments have received the technological and software support necessary to develop telecommunications systems. Countries whose governments have maintained independence from, or exercised control over, U.S. business interests have, by and large, not fared as well.

In Central America, television has increased largely in recent years due to the use of satellite and cable distribution. It is estimated that there is the equivalent of one TV set for every family in Latin America—that is, both Central America and South America. However, in a number of countries, in great part because of the lack of monetary resources for receivers as well as for distribution, radio remains the principal electronic medium, reaching the public through terrestrial signals that, with sufficient power, can cover an entire country. Countries continuing to rely principally on radio include Guatemala, El Salvador, Honduras, and Panama. Radio signals are extended through microwave systems.

Television distribution systems operate principally through terrestrial stations, many of which receive signals through satellite reception of U.S. programming. In the early 1990s cable began making heavy inroads. Some countries with good economies and tourism, such as Belize and Costa Rica, have moved forward rapidly with satellite-received, cable-distributed TV. Some countries that use the media especially for educational and political unification purposes, such as Cuba, also have developed strong TV systems. All countries have earth station links with the multinational International Telecommunications Satellite Organization (INTELSAT), and all use the United States' NTSC color system.

Mexico

Mexico, with over 90 million people, is the largest in population and area of the lower North American countries. Its extensive system of radio and television stations reaches not only all parts of Mexico, but neighboring countries as well. Therefore, it influences the cultures of other Central American nations through the programs it produces and distributes. Mexico has domestic satellite terminals and a number of earth receiving stations that pick up foreign TV programs; as with other countries in the region, these programs are principally from the United States. In 1992 Mexico had one government-operated and one privately owned television network, each with multiple channels, and a number of additional independently owned stations. Cable offers additional programming, but thus far its cost has limited its use principally to the relatively few who can afford it and to resort hotels. In 1993 the TV-set-to-persons ratio was 1:6.6. However, as cable competition increases, technology improves, and subscriber fees go down, cable coverage is expected to increase.

In Mexico, as elsewhere in Latin America, reasonable rates and expansion of services to the economically deprived depend on the politics of the parties in power. In the mid-1990s monopolies in Latin America were increasing, with many ownership mergers and transfers. This control over programming was in part obviated by growing satellite operations, which promised greater access to reception for some people, depending on the cost and availability of home receive-dishes. PANAMSAT (Pan American Satellite Corporation) began plans in 1994 for a DBS operation for Mexico and elsewhere in Latin America.

Radio systems, like television, are operated by both government and private entities. The ratio of radio sets in Mexico in 1993 was one set per 5.1 persons. Radio programming is similar to that of the United States in the age of television—principally music and news.

Some public-interest factors are programmed into the government network's channels, but all the other stations reflect the conservative, status-quo viewpoints of their wealthy owners. The government controls and regulates telecommunications under the Secretary of Communication and Transportation and through its General Directorate of Radio, Television and Cinema.

The other Central American countries by and large have similar systems and radio and television set penetration. As noted above, telecommunications use in most of the Central American countries is limited by their generally third-world economic conditions. With few exceptions, the percentages of populations with radio and TV sets ranges from 5 to 30 percent. Because of the high rate of illiteracy in most third-world countries, including Central and South America and most of Africa and Asia, television plays a major role in communicating information, as well as entertainment. It also, as one might assume, plays a significant role in influencing the purchasing habits of a large number of uneducated or undereducated viewers, and uses, by and large, the U.S. approach to media advertising.

Guatemala

In Guatemala, the second largest country in the region with 10 million people, widespread poverty (especially among the Mayan peoples), and a lack of electricity in many rural areas of the country, less than 10 percent of the population own radios and only 5 percent own television sets. But keep in mind that listening and viewing groups are usually larger than in the United States, and the number of people reached by one set at any given time may be four or five times the percentage of sets for the population. The ratios are 1:18 for TV sets and 1:22 for radios. This holds true for other countries as well, so it is important to look at the percent-per-population or ratio figures as starting points only.

As in Mexico and other countries, resort hotels in Guatemala have cable in order to give their guests entertainment and news programs, principally from the United States. Guatemala has a General Director of Broadcasting and National Television, under the Ministry of Transportation and Public Works, overseeing media operations.

El Salvador

Support by the United States of its dictatorship enabled El Salvador to develop its telecommunications facilities to a greater extent than many Latin American countries, partly to maintain propagandistic control over its people. In 1993 about 40 percent of El Salvador's 6 million people had radios (1:2.6) and about 8 percent TV sets (1:12). Radio is still the key medium, but television and cable were growing rapidly in the early 1990s.

Government-owned TV stations concentrate on educational programming and news, with the privately owned commercial operations programming shows principally from the United States and Spanish-speaking countries outside El Salvador. Both the government and the private sector have a hand in controlling broadcasting in El Salvador: the government's National Administration of Telecommunications, and the Salvadorean Association of Broadcasting Contractors, the latter somewhat similar to the U.S. trade organization the National Association of Broadcasters. As a dictatorship that kept tight control of its people through military control, including the infamous death squads and sanctioned torture, El Salvador likewise has kept tight control over programming content, at the same time cooperating with the U.S. investors who have developed the country's communications technology, programming, and advertising revenue.

Honduras

Honduras, about the same size as El Salvador with over 5 million people, in 1993 had about the same penetration of radio, about 40 percent of the population (1:2.4), but only about 4 percent (1:24) had television sets. There are national networks and regional and local stations. Although cable is available in Honduras, the economic conditions have made connection a slow process. A government- authorized-and-controlled private group, the Honduran Contractors of Television, is the principal coordinating media body. Almost all radio and TV stations are privately owned, but through its federal agency,

HONDUTEL, Honduras maintains close regulatory control over the private as well as the governmental stations.

Other Central American Nations

Some countries, like Nicaragua and Cuba, have had difficulties in developing their telecommunications systems. Both overthrew dictatorships to establish socialist governments and both suffered as a result of United States-led sanctions, such as embargoes and military assaults, including an attempted U.S. invasion of Cuba, and the United States-backed Contra forces in Nicaragua. Sanctions also included a ban on telecommunications investments and technology. In Nicaragua only about 25 percent of the population owned radio sets (1:4.3), and 5 percent (1:18) owned TV sets in 1993. The state-owned television system broadcasts cultural and entertainment programs on two national channels. Since the change in Nicaragua's governing party, the United States has eased sanctions against the country and more investors are now assisting the growth of the country's telecommunications industry. However, as this is written, sanctions continue in force against Cuba, which maintains a tightly controlled system devoted principally to cultural and educational programs. Cuban government head Fidel Castro has long recognized the importance of TV, as well as radio, and has made television a key political medium in the country. In Cuba in 1993 one in 5 persons had a television set (almost every household), and 33 percent of the population had radios.

An example of a country independent of U.S. control, and an anomaly in a region beset by poverty and revolution, is Costa Rica. As an independent, democratic country, Costa Rica has been able to develop its telecommunications system within a good working economy. One government-run station provides primarily cultural and educational programming, and five private stations principally broadcast entertainment programs imported from the United States and from Spanish-speaking countries. Telecommunications is regulated by a government agency, the National Control of Radio. Its television-to-person ratio was 1:4.9 and radio-to-person ratio was 1:11 in 1993. Panama is an example of a Central American country that recently has attempted to exercise more political independence, after many years under U.S. domination. But like many others it cannot grow without economic assistance from the United States, for its telecommunications as well as other areas. About 20 percent of its 2.5 million people have radios, and about 10 percent TV sets. Panama also brings in foreign programming via satellite. The government controls broadcasting through its National Directorate of Media and Social Communication. While radio is the principal medium in the country as a whole, television is dominant in the capital, Panama City, where 90 percent of the households have TV sets and where the use of cable is growing.

It is worth mentioning one small country in Central America, Belize. Belize broadcasts in English as well as Spanish, recognizing its role as a world tourist center. Its economy is good and although it is in political conflict with neighboring countries concerning its borders, it otherwise remains relatively independent of outside influence. It has at least one radio per household, and most of the households have TV sets, which receive programming by satellite, principally from the United States, distributed by two private TV stations. Radio as well as television is dependent on programs from other countries, especially other Central American countries.

An overview of the southern part of North America should include recognition of the twenty-six Caribbean Islands. In most of these islands, telecommunications systems are still in the developmental stage. While the British were dominant in the area for many years, other countries also used the islands for various economic and political purposes, and programs in the islands can be heard in English, Dutch, French, Spanish, Creole, and Papiamento, among other tongues. Most of the islands are considered third-world countries, existing in various stages of poverty. Political unrest in the islands is monitored carefully by the United States, which maintains a strong military as well as economic presence in the area. The various countries' political histories, as well as economic dependence, preclude much origination of programming, with the exception of music, talk, and local news shows. Programming consists principally of foreign imports. U.S. programming is especially in demand, with much of it pirated from communications satellites that have footprints (areas covered by the satellite signals) over the region. It should be noted, however, that while U.S. programming is popular, its commercialism has had a serious negative impact on the attitudes and behavior of some of its viewers in the region. One bizarre situation occurred on the island of St. Kitt's. The government constructed new athletic facilities, featuring a basketball court, for its youth. However, it found that there was virtually no use made of the basketball court. The solution to the mystery was that the young people, exposed to the commercials on the imported U.S. programs, felt there was no point in going out onto a basketball court unless they had the "right" and "acceptable" athletic shoes they had seen so frequently advertised by top U.S. basketball players—athletic shoes that most people on this impoverished island simply could not afford to buy.

CARIBVISION began to make a significant contribution in the mid-1990s, delivering through satellite programs especially designed for or of interest to the Caribbean nations.

SOUTH AMERICA

With few exceptions, such as the larger nations of Argentina and Brazil, tele-communications in South America is similar to that in Central America. While the U.S. influence is not as pervasive, the depressed and, in many cases, impoverished economic conditions are. Although the advertising market is growing, investments and technology have a long way to go. Radio penetration is high in South American countries, with people in urban areas and in smaller cities and towns having that medium available through household sets or public transmission in market places and shops. Television stations are concentrated in major cities, with their signals rarely reaching rural areas of the countries.

Countries that have had close economic ties with the United States or European countries have, as might be expected, the most advanced systems in structure and operation. South American countries have an advantage over their northern neighbors: picture quality. While North and Central American countries use the United States' inferior NTSC standard, most South American countries use the higher-quality German color system, PAL (used by more countries than any other format). South American countries receive signals from a number of satellites, some countries from more than one. All countries in the region, with the exception of Guyana and Suriname, are members of INTELSAT. A number of countries belong to the International Maritime Satellite Organization (INMARSAT), which provides global links for safety and disaster communications. PANAMSAT began broadcasting in 1988 to all countries in South America that wish to receive its signals. CONDOR is another regional cooperative satellite project slated for operation in 2005.

Brazil

Brazil has the most sophisticated tele-communications network in South America, with a ratio of one TV set for every 5 of its 160 million people—a high percentage of all households. It has the world's fourth largest TV network, the privately owned Rede Globo. Only the United States' three leading networks, ABC, CBS, and NBC, are larger (although in terms of distance and numbers of viewers, Russia, and China, respectively, might be said to have the largest television networks). Brazil was a leader in the region and in the world in utilizing television to forward national and public interests. In the early 1970s it was among the first countries planning to use satellites to reach parts of the country inaccessible through terrestrial signal distribution. Brazil's state-owned network provides mainly educational, cultural, and informational programs, with no commercial advertising. In the 1970s it used television to open educational and cultural doors to vast numbers of its population who were previously isolated from the mainstream of society. It is estimated that in a period of just a few years over one million illiterate Brazilians learned enough through television to become viable in the employment marketplace.

Brazil is becoming more and more involved in the international market, both in exporting its own programs and in importing others' programs. In the early 1990s it had program exchange agreements with eighty-one countries. It uses not only INTELSAT, but its own BRASILSAT service. PANAMSAT and DBS use is also growing.

Its nationwide radio system is equally sophisticated, with virtually every household having at least one radio set, with a ratio of one set per 2.5 persons.

Argentina

Argentina, with 35 million people, has TV sets in almost every household, one set for every 4 people in the country. In addition, there is one radio for every person living in Argentina. What began as state-run television had almost a complete turnover to private ownership in the early 1990s, with the government retaining one network and private parties operating four others. Cable has made strong inroads in Buenos Aires and, in the early 1990s, was expanding in a number of other areas of the nation.

Argentina, like Brazil, uses INTELSAT and expected to begin operation of its own domestic satellite system, NAHUEL, in 1995.

Uruguay

Brazil's neighbor, Uruguay, a poorer nation, also has one TV set for every 4 people of its population of 3 million, and a 1:1 ratio of radio sets. In the early 1990s Uruguay instituted a nationwide microwave network to expand its coverage, and was in the planning stage for the introduction of cable. It already has two INTELSAT earth stations. Both countries use PAL, facilitating across-the-border program exchange.

Bolivia

Bolivia, one of the poorest South American countries, had a ratio of one TV set per 16 people in 1993, which indicates that about one-third of Bolivia's 7 million people see television on a regular basis. The country is served by one state-owned network, using microwave to transmit to the public through thirty regional stations. Cable was growing slowly in the early 1990s, limited then to the nation's principal city, La Paz, and carrying Brazil's Globo TV and several U.S. channels. Its radio set ratio is one per 1.8 persons, and as the 1990s progress, Bolivia is attempting to expand its radio service through nationwide relays, and its international programs capacity.

Paraguay and Peru

Paraguay's 5 million people are served by two private TV networks for its 10 percent of the population (almost half the households) with TV sets. Cable was expected to be introduced into the capital city, Asuncion, in the early 1990s. Microwave relays are able to extend TV and radio signals throughout most of the country. Paraguay's radio ratio is 1:5. Peru has one state-owned and seven private networks, all based in its capital, Lima. While radio remained dormant in the rural areas, TV grew rapidly in the early 1990s, with PANAMSAT,

INTELSAT, and a dozen domestic ground stations, principally servicing Lima. The nation's 23 million people receive, in addition to Peruvian programming, programs from the United States, Mexico, and Argentina. In 1993 the TV set ratio was one per eleven persons and radio, one per five persons.

Ecuador

Ecuador, in the early 1990s, served its 11 million people with three private networks from two sites, Quito and Guayaquil. There was one radio for every 3 people—a high percentage translated into radio homes, and one TV set for every 17 persons. Ecuador is one of the several countries where cable has begun to make serious inroads into the traditional broadcasting system, with one cable system already providing forty channels in the early 1990s.

Venezuela

Venezuela, better off economically than most other South American countries and with a relatively stable government, has a sophisticated system for its 21 million people, 20 percent of whom owned TV sets in 1993: two state-owned, three private, two regional, and one pay-TV network. About one in two people owned a radio set. Cable TV has begun to expand in the capital, Caracas, carrying mostly U.S. channels.

Colombia

The same percentage of Colombia's population of 34 million, 20 percent, owns television sets. Colombia has one state-run network, with three channels serving the entire country from Bogota, with sixteen hours of programming a day. In addition to its own productions, Colombia TV imports programs from Brazil, Canada, Great Britain, Mexico, the United States, and Venezuela. Despite being a small country whose media in recent years have been caught in crossfire pressure from the government and the drug cartels, Colombia began early

to use television for educational and cultural purposes. In the early 1960s it was the first country in South America to develop TV for instructional use, with the help of a U.S. Peace Corps team. Radio is pervasive in Colombia, with one radio for every 7 persons.

Chile

Chile, with a population of 14 million, has one of the more interesting broadcasting systems. Its single state-owned radio and TV networks are supplemented by stations operated by the country's universities. Some of these TV stations are distributed nationally through satellite. Chile also has an expanding cable service, with three systems operating in its capital, Santiago, and rapidly expanding to other parts of the country. In 1993, Chile had a one-third population saturation rate for radio and a one-fourth rate for television sets.

Guyana

Guyana's broadcasting system is representative of those countries whose entire infrastructure quality, including education, the arts, and communications, varies with the political philosophy in control at any given time. Over the years, Guyana has moved between open communications and high literacy under more democratic rulers to poor education, low literacy, and a repressed communications industry under restrictive rulers. Under its new president in the late 1980s and early 1990s Guyana began a liberalization of its media policies, adding private stations to the formerly government-controlled system. Much of its TV programming, however, continues to be rebroadcasts of programs from the United States. While television is only now, in the 1990s, beginning to grow, almost every household has a radio set.

Suriname and French Guiana

Two of the smallest nations in the western hemisphere, Suriname, with a population of less than a half million, and a 10 percent ratio of TV sets, and French Guiana, with fewer than 100,000 people and a TV set ratio of about 7 percent, orient their TV systems to their European parent nations. French Guiana uses France's SECAM color system format to receive broadcasts from France's major TV networks. Suriname's two networks, one government and the other private, provide programming in Dutch and in English, primarily imported from the Netherlands, the United Kingdom, and the United States. Suriname has total radio saturation, with one set per 1.5 persons.

EUROPE

While no single European country is probably as globally influential economically, technologically, and in programming distribution as the United States, some European countries are at least as sophisticated in program production artistry and have become centers for worldwide news gathering and dissemination. In fact, the world's focal point for satellite news gathering and distribution is not New York, but London. Further, some areas of northern and western Europe are more densely covered with broadcasting stations than any other sections in the world. With so many countries squeezed into a small geographical area, there is not enough spectrum space for all the stations needed to provide programming to all the different ethnic, national, and special-interest groups. The spectrum in many areas is overcrowded, limiting signal strength or resulting in signal interference. The close proximity of countries also results in cross-border reception and in unauthorized station operation in some countries. For example, probably the most popular entertainment channel in the Netherlands, with Dutch star performers and national personalities, is not licensed by the Dutch government but is headquar-

tered in and distributes its signal from Luxembourg.

Many countries welcome programming from other nations and have entered into agreements for program exchange. With increased trade through the Common Market, the lifting of travel restraints by the European Union, and economic conditions forcing workers across borders to seek jobs, many countries have nationals from other countries who want to see programming from their homelands. A number of satellites serve Europe for program exchange, ranging from INTELSAT globally to EUTELSAT over Europe to BSKYB from the United Kingdom.

It should be noted that the programs seen in Europe are of better picture quality than those transmitted in the United States. European countries have either PAL or SECAM systems, with greater linage and therefore better pictures than the U.S. NTSC system.

In all European countries, including the United Kingdom, the principal source of funding for broadcasting is through license fees charged to radio and TV set owners. Where private stations have been authorized to operate commercially, and where ads have been allowed (albeit with limited time and in designated segments) on government or public-operated channels, advertising provides from full to partial revenue. Almost all European countries have moved from systems initially under government control and operating usually as a public service, to a combination of state-owned or -supported systems and privately owned systems. Those that still have only government channels, such as Austria and Portugal, were in the early 1990s looking into the authorization of private channels.

An unanticipated by-product of privatization has been critical competition for advertising among the commercial channels. With an increase in the number of private radio and TV stations, the availability of advertising for any given channel decreases. In some countries, therefore, private as well as public systems are dependent, at least to some degree, on government allocation of funds from license fees. Some countries, such as Italy, take advantage of this economic dependency to favor stations that reflect the ideology of the government in power. The cooperation growing among national telecommunications systems through the EU has resulted in regional advertising, increasing the commercial time and revenue for many stations in the region (especially those in Austria, Greece, Italy, Portugal, and Spain) that belong to the Common Market. To avoid overcommercialization, the EU has adopted limitations on advertising content and volume.

Communications satellite distribution testing began in 1978 when twelve countries experimented with the delivery of television signals to much of Europe through Orbital Test Satellites (OTS) launched by the European Space Agency (ESA). The European Communication Satellite (ECS) was added in 1983. These satellites provide programming to terrestrial stations and cable systems. All European countries have satellite links, distributing international programs from the STAR and SKY satellite channels, including BBC and CNN news programs, on cable.

Cable has become a principal means of distribution in many countries, but disproportionately in western Europe. Belgium and the Netherlands, for example, are virtually completely wired. In southern Europe, however, in the early 1990s only Spain, Austria, and Switzerland had cable, with the latter two authorizing national cable franchises (Figure 2-1).

In 1995 two audio phenomena hit Europe: mulitchannel music with CD-quality stereo through satellite, and pay cable radio. Astra Digital Radio (ADR) planned to distribute more than 700 radio and pay-audio channels through its four satellites. Anyone with a small satellite dish in Europe could receive the signals. At the same time, Eutelsat was preparing to offer digital pay-radio services through satellite to cable systems throughout the continent.

Eastern Europe is an anomaly. Even as this is written, four years after the "velvet revolution" and the dissolution of the Soviet Union, it is not clear in which direction telecommunications is going in any given country. For example, while television and radio stations became independent in 1991 following the first democratic elections, in 1993, when Russian president Boris Yeltsin dissolved the Parliament and seized dicta-

FIGURE 2-1
Cable use, while high in some European countries, is almost nonexistent in others. Courtesy Canal +.

CABLE PENETRATION IN EUROPE AT END 1993

	in %
Netherlands	90
Scandinavia	50
Germany	42
France	6
United Kingdom	3
Spain	1
Italy	0
Europe	24
USA	59

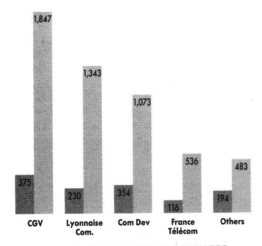

BREAKDOWN OF CABLE SUBSCRIBERS AND HOMES PASSED BY CABLE OPERATOR IN FRANCE AT END 1993
(in thousands)

■ Cable subscribers
▨ Homes passed

torial powers, he took control of the media as well, returning them to being government propaganda machines, as they had been under the Soviet regimes. On the other hand, in Poland the media became independent of government control, except for being careful to maintain policies that would guarantee continued government funding. However, the re-emergence of a strong religious fervor in Poland has resulted in the media being unofficially and tacitly, but effectively, controlled in content and coverage by the Catholic Church. In the early 1990s other eastern European countries were in the process of rewriting their communications laws, and some experimented with new systems, changing them as they went along. While several new systems have emerged in the mid-1990s, it is impossible to predict that any of them will continue in their present forms in the future.

United Kingdom

The British Broadcasting Company (BBC) has served as a world model since its establishment in 1922. Whereas the United States permitted signal chaos during the early years of radio, the British government established the BBC as the single broadcasting entity. The government levied license fees and what was called a "poll" tax on households with radios as the source of funding, and chartered the BBC as a non-government, independently run organization. The BBC's programming was oriented principally toward information and education. While the United States led the world in technological distribution, the British were consistently first in innovation, such as the first test of television in 1926 and demonstrations of high-definition TV in 1936. After World War II, the BBC was reorganized into three radio networks, serving "highbrow," "middlebrow," and "lowbrow" tastes. Over the years the BBC monopoly has slowly been eroded by the licensing of independent television systems that operate much like the American commercial networks and stations. Cable came to the United Kingdom in 1983 and in the early 1990s was served by twelve satellite channels feeding cable head-ends (the receive-points for programs to be distributed by wire). Private satellite-receive dishes, receiving satellite signals directly, were authorized by the government in 1985. By the 1990s about one-third of all British households had VCRs. The BBC World Service broadcasts news, public affairs, education, and entertainment programs internationally. The Radio Authority is the licensing and regulatory body for the independent commercial radio channels, relatively recent competition for the BBC radio channels. The Television Act of 1954 created independent television as competition for BBC-TV. The Independent Television Commission (ITC) is the public agency responsible for licensing and regulating commercial TV in the United Kingdom. The United Kingdom launched Europe's first satellite service in 1982; that later merged with a second satellite service to become British Sky Broadcasting (BSkyB), which in the mid-1990s was providing multichannel services all over Europe. In 1983 cable began in the United Kingdom with the awarding of eleven pilot franchises, and in the early 1990s was expanding rapidly, with new franchises granted by the ITC for various parts of the country. The UK's 60 million population has a high density of television sets (one per 3 persons) and radios (one per person).

Ireland

The Republic of Ireland has a system similar to that of the BBC. It began with the Wireless Telegraphy Act of 1926 and with one station, Dublin Broadcasting, most of its programs coming from the BBC. In 1961 television began under the Irish Public Television Authority. Both television and radio in Ireland are mandated to focus their programming on Irish culture, nationalism, and education. Irish channels compete directly with four British channels and with a recent growth in VCRs. The development of satellite broadcasting in Ireland depends on the extent to which the economy will permit full utilization of the five European DBS channels assigned to the country. Ireland's population of 3.5 million in 1993 had a TV set ratio of 1:3.8 and a radio set ratio of 1:1.7.

Western Europe

Germany

Germany began radio programming in 1923 under the control of the federal Reichspost (PTT). Entertainment programs were paramount, with news and public affairs programming tightly controlled by the government. Television was introduced in 1935. Through the reign of the Third Reich, the Nazis used radio very effectively for propaganda purposes, combining it with print and film under the control of Joseph Goebbels. Following the end of World War II, the German broadcasting system was reorganized by the Allied powers. In the new German constitution of 1949 freedom of the press was guaranteed to both the print and electronic press media. In 1952 television was reestablished with a structure similar to that of the BBC, with license fees on set owners to fund both radio and television. With the heating up of the "cold war" and the construction of the Berlin Wall, East Germany's broadcasting system came under the control of the government, operating much like the other state-controlled systems in the Soviet sphere. West German broadcasting was divided into regions— south, north, west, and Berlin—with public-representative boards determining the operations and programming of each area. Despite much protest from the public broadcasting systems, cable was established in Germany in the mid-1980s, principally to give private entrepreneurs media outlets so they could profit from commercial advertising, severely limited on the public systems. Cable now constitutes a significant part

of telecommunications in united Germany, receiving programs at its headends from communications satellites, just as the older, terrestrial stations do. Germany launched its first direct-broadcast satellite in 1987, and the number of individually owned receive dishes is growing rapidly. Stations, cable, and individuals can arrange to get signals from many of the satellites, including British and French, that hover over Europe. Like most industrialized nations, Germany also has shortwave international stations, the Deutschlandfunk and the Deutsche Welle, the latter literally built into a snow-and-ice-covered Alp at the southern edge of the country. The Bundespost—equivalent to the Reichpost of the early days of German broadcasting—is the national licensing, regulatory, and technological development agency for all national telecommunications, including cable, while as noted earlier, the regional districts, or Länder, have similar jurisdiction over the regional public stations.

The difference in economic status of the former East and West Germanys is apparent in their set saturation in 1992, shortly after reunification. In both parts of this country of 80 million, almost 99 percent of households have radio and TV receivers. The ratios for the country as a whole were 1:2.6 for TVs and 1:23 for radios. However, the west has about twice as many TV homes with VCRs (almost 60 percent) and almost twice as many cable homes (about 40 percent). The reverse was true for satellite-equipped homes, with 15 percent of the homes in East Germany satellite-equipped—about three times the percentage in the west.

France

For many years France was an example of responsible, eclectic programming on government-operated channels. Radio began in the early 1920s and television in the mid-1930s, developing into a mixture of public and private stations. As in some other European countries, after

World War II, France nationalized all stations because of the private stations' collaboration with the German occupiers during the war. In the early 1990s, however, France once more reversed itself and began to sell off two of its three TV networks to private interests. A mixed radio system began a decade earlier in 1982, when the socialist government authorized cable and satellite communications and legitimized the many private stations that were operating in and around the country. Although privately owned radio and cable services were permitted, the government continued to own the transmission facilities. The operation (excepting the privately owned channels) of broadcasting is still under the Office of Radiodiffusion-Television Francaise (ORTF). The Audiovisual Communications Authority is the licensing and regulatory agency for all private stations. Another office, Telediffusion de France, supervises the technical facilities.

Three radio networks reach 99 percent of the population, one with news and general entertainment and the others with more specialized arts, education, and cultural materials. Television in France operates with little censorship of materials, and some programs would be considered too risqué for self-censored U.S. television. It offers a broad variety of entertainment, information, and education. As in other countries, the public systems are funded by license fees on TV and radio set owners.

There is considerable exchange of programming with neighboring nations, as might be expected, and technological cooperation for mutual telecommunications benefits. For example, in the 1970s France and Germany jointly launched two DBS satellites. France also operates an international shortwave service. In the mid-1990s advanced fiber-optic cable was reaching into almost every TV home in France under the direction of the Ministry of Posts and Communications, the licensing, regulatory, and technological development office for all of telecommunications in France.

France's 58 million people in 1993 had a TV set ratio of one per 2.6 persons and a radio set ratio of one per 1.1 persons.

Belgium

Belgium, one of France's many border neighbors, is not only related to France by political history, but by a common language for most of its people. Conversely, the Flemish-speaking segment of the population emphasizes the country's independence from France. It is not surprising that when television was introduced in 1953 the French-speaking part of Belgium selected the France SECAM system, and the Flemish-speaking part chose the PAL system used in most other European countries. It became necessary to build TV sets for use in Belgium that would accommodate both standards. Radio began in 1923 with a French-language station, Radio Belgique. The first public radio station was established in 1930 under the Belgisch Institute National de Radio (INR). Under the law setting up the station ("The Law of June 18, 1930"), the station was barred from being used for propaganda purposes by the government operators, specifying complete impartiality in news content and presentation. During the early 1930s a number of private commercial radio stations were authorized to go on the air. During World War II all station operations were suspended and the equipment was used by the German occupation forces to jam broadcasts from the BBC. Following the war most private stations were not permitted to go back on the air, leaving the government stations dominant. Two public TV systems were developed: Belgische Radio et Television (BRT) and, oriented to the French-speaking population, Radio et Television Belgische-France (RTBF). In 1993 the 10 million population had a TV set ratio of 1:4.2 and a radio ratio of 1:2.2.

Cable began to grow in the 1970s, with competing systems covering the country, and in the mid-1990s Belgium probably is the most intensively and extensively cabled country in the world. Neighboring countries broadcast over the cable systems; the Netherlands, for example, began pay-TV programming on Belgian cable in 1985.

The Netherlands

The Netherlands, which in 1919 was the first nation to put a regularly scheduled radio station on the air, has both the most democratic and the most complex telecommunications system in the world. Air time on television and radio is allotted to civic groups (labor, religious organizations, educational organizations, political parties) based on the group's number of members in relation to the country's population. Channel space is divided into pro-rata hours. These "Pillarization Principles" go as far back as the 1920s. Television began in 1951 as a noncommercial operation. Holland's Broadcasting Act of 1967 put the use of channels under the Netherlands Broadcasting Foundation (NOS); however, TV and radio station operation and transmission remained under Radio Nederland. The Netherlands is second only to Belgium in cable penetration, which developed strongly in the 1980s, with about 90 percent of the country wired. Holland exchanges programs with other countries via satellite, with principal distribution on cable. In 1993 Holland planned a second telecommunications network, to be operated by private cable, utilities, and railroad companies.

In the early 1990s the Netherlands divided the time on its three television networks among the major "pillarization" organizations. Perhaps the most popular TV channel, however, is not a Dutch station, but a Luxembourg station, RTL4/Eurosport, with live programs produced in Holland with Dutch performers, and the signal relayed to Luxembourg, where it is sent over the air back to Holland.

Radio has five networks serving the various pillarization groups. Competition is strong from two major radio stations, one beaming from Italy and the other, Sky Radio, from the United Kingdom. Radio and TV household penetration in the Netherlands is about 99 percent, with one TV set for every 3.2 persons and one radio set for every 1.2 persons in its population of 15 million. The city of Hilversum, near Amsterdam, is the Dutch broadcasting center, with even the studios of the Luxembourg station quartered there.

Luxembourg

Luxembourg's population of 400,000 has total home saturation of TV and radio, with ratios of 1:4 and 1:1.6, respectively. As in Belgium and other countries receiving signals from neighboring nations with different technical systems, sets can receive both SECAM and PAL modes. Radio-Tele-Luxembourg operated five TV channels in the early 1990s. Private companies own non-public stations, such as RTL4, that broadcast commercial programs to other countries as well as in Luxembourg. In addition, Radio Luxembourg broadcasts in the five languages that almost every Luxembourger knows and that can be received and understood in neighboring countries: English, German, French, Dutch, and the Luxembourg dialect.

Finland

Perhaps because of their geographical separation from other areas of Europe, the Scandinavian countries have developed somewhat different and protective telecommunications systems. Stressing public interest as part of their socialist economic and political orientations, Scandinavian countries have largely resisted the privatization of the media. They have also been especially concerned with the protection of their national cultures and arts and have attempted to keep out what they consider low-quality foreign entertainment, at the same time trying to develop programming on their government channels that can effectively compete with such programming. For example, YLE, the Finnish Broadcasting Company, has been attempting to retrain its writers and producers to develop programs that will maintain Finnish history and cultural traditions while also attracting audiences that might otherwise turn to satellite, cable, and private stations. As with other Scandinavian systems, Finland's broadcasting is governed by a board that is ostensibly representative of the interests of the public as a whole; in this case the board is composed proportionately of representatives of the political parties in the Finnish parliament. In 1990 the two Finnish public TV networks and their 40 stations were challenged by a newly authorized private network. The three YLE radio networks had competition from almost 60 local independent stations. Local cable stations also provide alternative programming. Because of Finland's large Swedish population, particularly on its west coast and border, there are both Finnish and Swedish broadcasts to its population of 5 million. The ratio of TV sets to persons is 1:2.7, and of radio sets, 1:1.

Sweden

Sweden's broadcast history is also similar to that of other Scandinavian countries. Radio began in 1925 under a government-established nonprofit organization, Sveriges Radio AB. For thirty years, Sweden had only one radio channel; then Parliament added a second and, in 1964, a third. The first television channel was authorized in 1956, and a second one in 1969. The Swedish Broadcasting Company was established to supervise all programming, including that on local radio stations throughout the country. In 1993 three operating entities were established, Sveriges Riksradio AB for national radio, Sveriges Television AB for national television, and Sveriges Utbildningsradio for international transmission. As in other European countries, funding comes from set owner license fees; in fact, advertising was permitted on only one public station. Almost every household in Swe-

den's almost 9 million population has both radio and TV, with a per person ratio of 1:2.4 for TVs and 1:1.2 for radios. Some 60 percent of all TV homes were receiving satellite signals through cable hookups. In 1992, legislation was being considered to establish private commercial radio stations, and by the mid-1990s privately run commercial stations were planning to beam their advertiser-supported signals throughout the country.

Denmark

Denmark has also reluctantly followed the trend toward a mixed public-private system; it was one of the last countries in Europe to hold out against advertising on TV. While the government operates the TV and radio networks, one of the TV networks operates as a commercial entity, although with limited advertising time. In addition, cable-distributed foreign satellite channels carry advertisements. Almost every household has both radio and TV, with network signals relayed to regional stations for distribution, and about 60 percent cable penetration as well. Of the slightly more than 5 million population, one in 2.7 has a TV set and one in 2.4 has a radio. Like other Scandinavian countries with limited resources, Denmark imports some of its programming, somewhat more than Sweden, about 25 percent of its program schedule, but somewhat less than Finland and Norway.

The growing exchange of programming among nations throughout Europe, through EU facilitation and continuing advances in satellite transmission and reception, creates a dichotomy of cultural aspirations. On one hand, each country is concerned with strengthening, or at least maintaining, its own culture; on the other hand, each recognizes the value of broadening its citizens' horizons with a better understanding of other cultures, thus improving the prospects of economic cooperation and peace.

Austria

Broadcast systems in southern Europe follow generally similar patterns. Austria is a good example. At first, Austrian stations were operated by the government for propaganda purposes. The party line had to be strictly adhered to. As governing parties or leaders changed, in Austria as in other countries, the directors-general or other heads of television and radio were also changed. However, by the 1960s, led by the print media, the country pushed for freer electronic media, and the Austrian parliament established the Osterreicher Rundfunk and Fernsehen (ORF) to oversee a new system of radio and television stations modeled after the BBC. ORF still runs telecommunications in Austria. Penetration of both radio and television is almost total. The principal transmission is from Vienna, with relay stations providing outreach to other parts of the country. However, Austria, like several other countries with Alps mountain ranges, has areas that cannot be reached by terrestrial or even cable signals. Nevertheless, one in 2.8 of the country's 8 million people has a TV set and one in 1.6 a radio.

Greece

State-owned-and-operated stations were the practice in Greece and Italy, too. The media were used to promote the respective government-in-power's concept of what was culturally appropriate and informationally sound for its citizens. Any deviation called for a change in directors. In Greece, for example, during the period of government shifts between parties of the left and the right in 1981–1989, the state-operated system, ERT, had thirteen different directors-general and sixteen different directors of news programming. Citizen pressure to eliminate such tight political control over media content resulted in the establishment of private ownership and operation of stations in 1988. Even so, the government's licensing authority continued to give it power; in 1989, the conservative government denied a license to a station whose socialist philosophy would probably have led it to support the opposition party. Further, some private stations that were unable to survive solely on advertising needed occasional fiscal help from their government, and therefore had to be sure that none of their program content could be found objectionable. Greece's population is just over 10 million, with a TV set ratio of 1:4.5 and a radio set ratio of 1:2.4.

Italy

Italy's introduction of private stations took place in 1976. However, although most broadcasting in Italy is privatized today, for years the government allocated channels and air time much in the manner of the Netherlands. Every major political party had its own channel or substantial time on a given channel. Other large and influential groups were similarly allotted air time. As in other countries, the commercial entertainment channels made the most headway with viewers and listeners. In Italy, entertainment moved into areas of freedom of expression that most countries still resist; after 11 P.M. nudity is permitted on Italian TV. In the early 1990s, for example, Italy's most popular TV show was a "strip" quiz show in which every time a participant failed to answer a question, he or she had to remove a piece of clothing, with contestants frequently reaching a state of complete nakedness. The Italian Broadcasting Company, RAI, continues to operate the public networks. Italy's media mogul, Silvio Berlusconi, has dominated Italian television; he owns all three private TV commercial networks. In addition, satellite signals are distributed by a growing number of cable systems and pay-TV. Public and private systems carry commercials, which are subject to time limits. Both radio and TV reach all parts of the country of 58 million people, with one TV set per 3.8 persons and one radio per 3.4 persons. While Italian radio today is principally entertainment, the tradition of use by the state is strong. Mussolini was once quoted as saying that if it were not for radio he would not have been able to achieve the control over the Italian people that he did. The election of a right-wing, fascist-supported parliament in Italy in 1994, with Berlusconi as prime minister, prompted fears that the Italian media would be seized once again for political manipulation. However, in early 1995 the volatile nature of Italian politics saw a new PM, and the feared takeover had not come to pass.

Spain

Spain and Switzerland are examples of countries that stress a combination of state-operated and private stations with multicultural programming. In Spain, the Basque population was suppressed for many years under the Franco dictatorship and was even forbidden to use its own language. Written records in Basque were largely erased, and the Basque heritage began to fade. Oral history survived, and following Franco's death the Basques were able to obtain their own radio and television channels, through which this oral history has been revived. While democratically opening the media to this part of its nation, in the 1990s Spain faced the irony of the media being used effectively by the Basques in their efforts to create a separate Basque state. Private television was authorized in 1982 and a Broadcasting Law of 1989 authorized local stations, cable, and hundreds of new radio stations. In the early 1990s, Spain had two public national TV networks operating under the government's Radio Television Espanola (RTVE) and three private networks, plus regional networks, with one in 2.6 persons of the country's 40 million population having a TV set. The RTVE radio network is carried by hundreds of local stations, with one in 3.4 persons having a radio. In late 1994 Spain authorized cable TV, with hundreds of systems wiring up much of the country.

Switzerland

Switzerland is multilingual, with much of the population speaking Italian, German, and French, in addition to Swiss dialects and English as a second language. The strongest concentration of multiple languages is, of course, in areas bordering the neighboring countries. Appropriately, the Swiss government operates three separate television-radio organizations, each broadcasting in one of the three principal European languages. Despite the difficulties of the terrain, the technological sophistication of the Swiss system provides excellent TV access, serving about 90 percent of the country's TV homes with TV and radio ratios of 1:2.9 and 1:2.6 in the 7 million population. All the broadcast systems are regulated by the Swiss Broadcasting Corporation, set up by the government along the lines of the BBC. Cable penetration is high in Switzerland, about 60 percent, with most TV homes able to receive satellite-distributed foreign programs.

Eastern Europe

The former so-called eastern bloc countries of Europe had comparable initial developments of their radio systems, but following World War II and the onset of the "cold war" (which lasted until the 1990s), their system developments differed from those of the west. In this chapter we provide an overview of eastern Europe as a whole. In subsequent chapters we provide more specific information on individual countries in eastern Europe.

While individual inventors in eastern Europe are associated with the early growth of radio (Russia, for example, continues to claim its scientist, Alexander Popoff, as the inventor of radio), the electronic media in those countries developed more slowly than in western Europe. Some reasons were the maintenance of strict government control for many years, the absence of private stations, and the post-World War II absolute ban on independent stations in most of eastern Europe until the early 1990s. Another reason was the economic condition of eastern European countries, which consistently lagged behind that of the western European nations.

Most European countries had regularly scheduled radio systems by the mid-1920s. To serve the various ethnic and language groups in most of these countries, many developed multiple-channel systems similar to that of Yugoslavia, designed to serve the nationalistic needs of many diverse and fragmented ethnic groups. Some did so long before Yugoslavia. But in some countries the multichannel systems were used to unify, not separate, the populations.

Rather than using the media to promote heterogeneity, World War II Yugoslav partisan leaders and later president Josip Broz—known as Marshall Tito—established three separate broadcast systems in the country's three major languages, Serbo-Croatian, Slovenian, and Macedonian, to maintain the autonomy of the three different cultures. Although all the stations were under the control of the government agency Jugoslavenska RadioTeleizija (JRT), they operated independently of and separate from one another, intensifying the nationalism in their coverage areas. The crises of the early 1990s in Yugoslavia illustrated how the media can be used for division and war instead of for understanding and peace.

The Soviet Union recognized the power of radio early, and shortly after the revolution of 1917, the new government used radio to send the new economic, political, and social messages of Communism to all of the people. It successfully used radio to solidify its base of support through an extensive and intensive government broadcasting system, emphasizing education, culture, and news. While setting standards for propagandistic use, Soviet broadcasting also set standards for high-quality artistic programming and expert use of radio's (and, later, television's) aesthetics.

Rebuilding telecommunications, including broadcast systems, destroyed during World War II was a slow process. After its armed forces liberated eastern European countries from the Nazi occupation forces, the Soviet Union undertook to reeducate the people of eastern Europe about the principles of Communism. This resulted in the development of extensive radio and television facilities throughout most of eastern Europe by the mid-1950s. The Soviet Union provided the principal technological and economic resources necessary to develop the first TV station in a given country and, later, a second or third. All stations were controlled by their respective governments. Programming in all countries was consistent with the communications policies established in Moscow. News and public affairs emphasized the policies and achievements of the Communist state. Cultural and arts programs reflected the purposes of socialism. Programming for young people was principally educational, stressing political, economic, and social, as well as academic, themes. Because of the rather narrow focus of programming, centralized bureaucratic control, and the lack of competition to generate viewer interest, programming techniques in eastern Europe were pedestrian, even though program content itself focused more on the arts than did any other region in the world.

The USSR linked the eastern bloc countries through satellite, its INTERSPUTNIK vying with INTELSAT globally and especially serving the countries of eastern Europe. In 1986 the Soviet Union inaugurated DBS service to all countries in its orbit. In the 1990s, with the dissolution of the Soviet Union and the opportunity for greater localization of media operations and programming, more and more countries and stations have begun creating independent structures and operations. However, various factors have prevented the kinds of restructuring and innovative programming desired and anticipated. One reason is instability (i.e., initial independence of broadcast stations in the new Russia was followed by dictatorial control of the stations by Russian president Yeltsin). Another is rapidly changing political control in many countries. Exploitative nationalism in some countries resulted in closer control of the media. Lack of funds prevented the upgrading of systems to state-of-the-art technology. Inadequate funds and facilities limited the retraining of writers, producers, and other media personnel.

Nevertheless, in the 1990s and even earlier where permitted, U.S. entertainment programming delivered via satellite became very popular. Prime-time soap operas like "Dallas" became the favorite programs in a number of eastern European countries, including Russia, as they had in the United States. Music videos and international sports, received principally through satellite channels such as Eurosport, also became very popular.

DBS is growing in eastern Europe, although the economies are not yet strong enough for many people to afford the cost. Satellite-receive dishes are springing up throughout eastern Europe, especially in the Czech Republic, Hungary, and Poland—three countries in the region with the fastest- developing economies. Part of the DBS growth in these countries comes through illegal dishes—that is, video piracy.

The popularity of western programs and the lack of commercial advertising in eastern European media for many years has made the region ripe for western entrepreneurs. Despite the efforts of literally hundreds of foreign investors, eastern European countries have thus far resisted any large-scale introduction of privately owned stations. Resistance to the funding available through commercial advertising is not as strong. U.S. cable companies have begun making inroads in eastern European countries, with an eye toward the thus-far untapped potential for national cable systems.

Operations have varied from country to country. The German Democratic Republic (East Germany) had the most extensive system outside of Russia, covering the entire nation with virtually non-stop programming. Romania, on the other hand, one of the poorest of the east European countries, has few channels, reaching relatively few receivers. Lithuania is another example of the trend in eastern European telecommunications since the dissolution of the Soviet Union. With a radio and television set in almost every home, Lithuania has coverage not only from its own state and private stations, but continues to receive signals from Russia. In addition to national networks, Lithuania has a large number of regional stations and relay transmitters.

Bulgaria's recent history of telecommunications development is illustrative of what has been happening in other eastern European countries following the dissolution of the Soviet Union. Bulgaria's broadcasting system has been operated by the government, closely resembling the structure in the Soviet Union itself. The state's Committee for Radio and Television controlled the media. Radio and later, in the 1970s, television were used primarily to inform and educate, not to entertain. As Bulgaria moved toward a market economy in the early 1990s, it concomitantly began to relax its control over broadcasting as well. As early as 1991 it began to include new laws for broadcasting in the rewriting of its constitution. The gradual privatization of many areas of the economy included broadcasting, and in 1991 ten private companies were designated to operate the national systems of communications, under government regulation somewhat similar to that in the United States and some countries in western Europe. Privatization of the media included the introduction of advertising, which began to account for a significant part of broadcasting budgets. Reflecting the economic and political trends, broadcast programming too began to reflect that of the West more and more, including prime-time soap operas, sitcoms, and dramas. While no two eastern European countries have followed identical paths, Bulgaria's approach does parallel the general changes and trends in the other former Soviet bloc nations in the region.

The number of television sets is remarkably high in eastern Europe, considering its relatively low economic status compared to countries in western Europe. Population and TV set and radio ratios per prson in the early 1990s for some eastern European nations were as follows:

Country	Population	TVs	Radios
Bulgaria	9 million	1:3.9	1:2.5
Czech Republic	10.5 million	1:2.7	1:3.3
Estonia	1.6 million	1:2.7	1:1.7
Latvia	3.8 million	1:2.9	1:1.2
Poland	39 million	1:3.9	1:3.6
Romania	23 million	1:6	1:7.3
Lithuania	10.5 million	1:2.7	1:3.3

Russia

The Soviet Union was the first country to distribute its programming via satellite, beginning in 1965. Satellite transmission enabled the central government station in Moscow to reach over 70 percent of the population in all of the Soviet republics, thus reinforcing its message of unity through its program content. Each republic had its own system that operated independently but was under the direction of appointees from the central government. Capitals of the various republics and larger cities had their own stations, with the largest, such as Moscow and Leningrad, having three TV stations. Now each new independent country and each large city has taken control of its own stations. While there is no central supervision from Moscow, in most of the countries the same kind of censorship, propaganda use, and content determination by the new independent and individual governments remain. In 1991 Russia began to accept applications for private stations, including one from a co-venture between the United States' Turner Broadcasting and the Moscow Independent Broadcasting Company. In 1995 Russia was in the process of replacing the state-run national television system, Ostankino, with a new system, Russian Public Television, part public and part privately owned.

Russia's vast territory requires creative distribution of broadcasting signals. It has a population of over 150 million and, stretching from eastern Europe across northern Asia to the Pacific Ocean, is geographically the largest country in the world. It has excellent saturation of broadcast receivers, one TV set per 3.2 persons and one radio per 1.5 persons. All technical facilities are under the jurisdiction of the Ministry of Communications. Russia combines various means of distribution, putting together the various elements like a jigsaw puzzle to reach across its eleven time zones. In the early 1990s it was using a half-dozen satellites, about 500 terrestrial transmitters, and more than 3,000 relay transmitters. In addition to hundreds of radio stations throughout the country, Russia has an estimated 35,000 wired radio systems that provide signals to virtually every home in the nation. Also, like other large nations with high political visibility and stakes on the global scene, Russia operates an external broadcasting service that in many respects is more extensive than those of the BBC and Voice of America, with regular listeners even in small towns in various regions of the United States. One commentator for Radio Moscow (the shortwave service), Joe Adamov, told us that once when he was visiting in the midwest of the United States he heard a comment on Russia over a local radio call-in show that he felt compelled to answer. As soon as he was connected and his voice was on the air, he was immediately recognized by the talk-show host as the Radio Moscow commentator.

The penetration of television and radio sets in the former Soviet Union is illustrated in the following early 1990s figures for two former Soviet Republics: Belarus has a population of over 10 million, with one TV set per 3.3 persons and one radio per 1.3 persons; the Ukraine, second only to Russian in size in the region, has over 52 million people and ratios of one TV set per 3 persons and one radio per 4 persons. Poorer countries like Turkmenistan and Uzbekistan had, for example, only one TV set per 6 persons.

AFRICA

Telecommunications in Africa is a panoply of contrasts. While some countries that were occupied by colonial powers for many years have sophisticated systems, others that were similarly occupied were left with virtually no telecommunications capabilities. While some countries are extremely poor and cannot afford even comprehensive radio (much less television) systems, others, in particular the Gulf States, have used their oil riches to bring the latest state-of-the-art developments to their countries, including satellite transmitting and receiving stations. But even in the Gulf States there is a paucity of programming because of censorship for political and religious reasons; thus their publics get little more (and sometimes less) variety of programming than some of their impoverished neighbors.

Most of the countries' systems were developed along the lines of the British system, in part because Britain was the occupying colonial power, in part because the United Kingdom had the most advanced system of any colonial power and thus provided the best model to copy. Exceptions to the British system are those countries colonized by France, such as Lebanon and Syria. After these countries became independent, they retained the same basic systems of telecommunications.

Another similarity is that after independence, all of the countries used their telecommunications systems to maintain control over the people through nationalistic education and propaganda.

In some countries, television was introduced on the basis of an event that the people of the country wanted to see and that the government wanted to promote, such as the Olympics or the visit of a world dignitary.

There are some strong national and many regional differences among the thirty-four countries of Africa, although most of Africa can be classified as third world, with the attendant scarcity of resources required for telecommunications sophistication and with the need to use communications systems to maintain political control in the face of continuing economic crises. In the western part of Africa, almost all of the countries are too poor to develop comprehensive television systems, and radio remains the principal broadcast medium. The scarcity and high cost of electricity gives battery-powered radio receivers a further advantage. Yet, as pointed out earlier, the cost of batteries creates an additional problem. Radio is of special value to governments in third-world countries because the generally high rate of illiteracy limits the use of newspapers and magazines as sources of information and persuasion, leaving radio as the principal means of affecting peoples' minds and emotions. Even in the somewhat better economic and educational climates of northern Africa, radio is the principal medium.

Even with the widespread availability of radio and its use to bypass the problem of vast illiteracy, the population's division into a great many tribal languages obviates the value of a national radio system in a number of countries.

Where there is television, the favored programs have been virtually the same as those introduced into entertainment-starved eastern Europe: U.S. prime-time soaps such as *Dallas* and *Santa Barbara*. In those parts of Africa where there were long-time colonial powers, most broadcasting is multilingual. In western Africa one finds French or English and each country's own traditional dialect. In northern Africa broadcasts are in Arabic, with some broadcasts in the second appropriate language, usually French or English. In the southern and eastern areas of Africa, with the exception of South Africa, television is scarce and radio has had relatively little penetration. Sub-Saharan Africa is clearly an impoverished area.

Categorizing African and Asian countries is difficult. In Africa the greatest differences are between the countries north of the Sahara, with their large Arab and Moslem populations, and the Black African countries south of the Sahara. Additionally, there are many differences between the Black African nations in the west and those in the east. Where we find differences significant enough to warrant separate sections, we will designate north, south, east, or west regions.

In addition, we will discuss the mideast countries as a subcategory of Africa, although a number of key nations, such as Iran, Iraq, Israel, Jordan, Lebanon, and Syria, are actually in Asia.

The Mideast

In the mideast, as in other parts of Africa and Asia, the first radio stations were developed by colonial powers. TV was introduced after the colonial powers left. These were national decisions made by the now-independent countries. Saudi Arabia, the United Arab Emirates (UAE), Qatar, Bahrain, Oman, Kuwait, and Yemen, frequently labeled the Gulf States (although Yemen is not on the Persian Gulf), had neither radio nor TV. These media were introduced in these countries later, all about the same time. All systems were introduced under the rubric of education—but this was found not to be a practical or realistic approach, and they soon shifted to, or added, entertainment.

By the early 1990s the oil money in several of the Gulf States made it possible for their citizens to have a high penetration of both radio and television sets. For example, population figures and TV and radio set ratios for these nations were as follows:

Country	Population	TVs	Radios
Bahrain	500,000	1:2.3	1:1.7
Kuwait	1.3 million	1:2.6	1:1.8
Oman	1.6 million	1:1.4	1:1.6
Qatar	500,000	1:2.5	1:2.5
Saudi Arabia	17 million	1:3.5	1:3.3
UAE	2.5 million	1:12	1:4.7

Conversely, Yemen, without oil riches, in the early 1990s had only one TV set per 38 persons and one radio per 35 persons for its population of over 10 million.

Three types of policies characterized the Arab states' media development and operation.

1. Egypt, Syria, Iraq, and Libya—after the colonial powers left they were formed as republics, not kingdoms—used radio and TV for political purposes. Their programming was all national, not imported. Lebanon, at one time considered the "Hong Kong of the mideast," is a separate case and attempted to be cosmopolitan and international in its programming.
2. After the colonial powers left, Jordan and Morocco became kingdoms, and their TV systems were used mostly for foreign entertainment.
3. The Gulf States stayed away from any controversial entertainment. Their systems were nonpolitical, concentrating on religion and tradition.

As time went on, countries that used the media heavily for political purposes generally moved to less political use, and countries that had used the media little for political purposes moved to greater use. After political upheaval in any given country, the trend in that country seemed to be toward more commercialization and entertainment. In the past decade or two, programming in mideast countries has grown more similar. They compete with each other and tend to make the programs more and more alike, as U.S. networks do, by cloning successful program types for short-term ratings gains.

Except for Lebanon, which had private stations, all the mideast countries had government-owned stations, usually within some ministry, most often the Ministry of Information, directly serving the propaganda purposes of the state. However, once countries were in control of their own media, innovation spread quickly, usually by one country trying to keep up with or take a step ahead of its neighbors—again, principally for propaganda purposes. For example, for many years Israel resisted the introduction of television. However, after Jordan introduced TV, Israel quickly followed. Trends also spread quickly. Most countries had one channel. When second channels were introduced, most countries followed the lead of using them for imported programming, principally English-language programs. Most mideast countries now have two TV channels; some have three or four. A new trend is toward regional stations. For example, in Egypt there is a Cairo station, an upper Egypt station, and a lower Egypt station, each oriented towards the culture of its region. But, to remain competitive in the country as a whole, there is a tendency for these stations to imitate each other.

Most TV stations are still linked to governments, principally because the governments want control over the news. In a number of countries, there is talk about private stations. Lebanon has some 50 private TV stations. Because TV is so expensive, some governments are being forced to consider privatization of stations and/or the introduction of commercial advertising. The Gulf countries, because they are rich, don't have to privatize or commercialize. Their command over content is strong because of religious control. They can't compete, however, with the programming of other countries that have entertainment. Some Gulf countries bring in less-cen-

sored materials underground. For example, there is the Middle East Broadcasting Centre, fonded in 1991 by a group of wealthy Saudi Arabian investors headed by the king's brother-in-law. It operates out of London, coming into Saudi Arabia through satellite with materials that can't be broadcast on Saudi Arabian stations. The government can disclaim any responsibility for the program content.

A principal reason for the Arab states—and other third-world countries—to import TV programs is given by Dr. Ziad Rifai of UNICEF, a Jordanian university professor: "They constantly have to feed the TV machine . . . the operator has invested in the hardware and must fill the time . . . in a country with limited production resources that can't produce all the programs that it needs, it is necessary to import foreign programs." The most available programs through satellite in Arab states are CNN and U.S. entertainment shows. The following table shows penetration into the Middle East and Far East by world TV satellite service:

Market	STAR TV (,000) Homes	% of TV Homes
China	30,363	22
India	7,278	25
Taiwan	2,376	46
Israel	621	49
Saudi Arabia	369	18
Hong Kong	331	21
Philippines	187	5
Korea	184	2
Thailand	143	2
UAE	117	29
Pakistan	77	4
Indonesia	50	*
Kuwait	31	11
Total	42,127	—

Source: Frank Small & Associates; courtesy STAR TV.
* Less than 1 percent

For both internal and external political purposes, countries tend to inflate the percentage of locally produced programming and underestimate the percentage of imported entertainment programming. Many so-called local shows include foreign segments. For example, a music/variety entertainment program locally produced may feature a western pop star.

Professor Rifai believes that there is increasing cooperation among Arab states. Arab states do not generally organize politically for international media clout, he says. However, individual states, such as Algeria, have been especially active in the International Telecommunications Union (ITU). While the Arab States Broadcasting Union sometimes coordinates action on major issues, Rifai states that usually there has been little coordination as an Arab bloc in relation to the rest of the world, but that this is changing.

There are differing views of the future of Arab telecommunications on the world stage. Tuma J. Hazou, Chief of External Relations for UNICEF in the Middle East and north Africa, told us that he doesn't see future close cooperation in broadcasting among nations of the Arab world except in news exchange—which he states is usually propaganda for the respective governments. Unlike some other media experts, he believes that the proliferation of stations will result in a glut, leading to a survival-of-the-best competition.

He categorizes telecommunications systems now in the Arab world into two types: ultra-right—devoted totally to blatant propaganda—and center-right. There is no really liberal or neutral type of service in the Arab world, he says. But Hazou believes this ultimately will begin to change. There are in 1994, for example, what he calls "slight superficial" changes happening in Jordan, Tunis, and some other countries—but they are not more than skin deep at this time.

Because most Arab leaders are not accountable to the people under the present governments and systems, the TV and radio programming does not necessarily serve the information needs of the people. Hazou thinks that at the present time there appears to be no major role for mideast broadcasting systems in the global picture—they will not rise to the level of the other systems because of political restraints. He says that most of the other world systems are established purely for commercial reasons—not for political or propaganda reasons—and Arab systems therefore are unable to find a key international role in mass communication.

Performers and producers in many developing countries try to copy western approaches—which do not necessarily fit their operations or audiences. Hazou says this is harmful to their systems. Frequently this is instigated not by TV professionals, but by bureaucrats in government who control the systems. Even the BBC is now being set up on a commercial basis. If this is true of the BBC, it is much more true of other regions. Relatively few countries do not now have commercial advertising on their systems. Hazou believes that the "best" systems will survive not necessarily on the basis of money or equipment. They will primarily need personnel, he says, who can decide what is a "good program." They will need experienced professionals. They will need to avoid political appointees in their broadcasting operations. "Give the dough to a baker to bake," Hazou says. He states that they need to experiment with new ideas.

Iraq

Iraq had one of the early radio stations in the region, a low-power station that started in 1935. In its early days it was used by King Ghazi to broadcast his anti-British, pro-Nazi views. It continued to broadcast throughout World War II, but not much was done with it after the war. However, following the Iraqi revolution in 1958, the government saw the station's value and expanded it. Iraq is credited with being the first of the Arab countries to recognize the importance of television. With British technology and assistance, TV came to Baghdad in 1956, and a second national channel was added in 1970. TV was at first used for political purposes; it broadcast programs of court trials against members of the old regime, encouraging the people to take out their anger against the old political system and to appreciate the new one. The broadcast system is under the control of the Ministry of Information, which practices rigid censorship and, frequently, distortion of the news. The government prevents the public from receiving across-the-border information by jamming foreign signals, just as it did for many months preceding the beginning of the Gulf War. During the Gulf War, its censorship—in order to galvanize public opinion in favor of its efforts—was probably as strong and complete as the U.S. military's censorship for the same reasons. In the early 1990s the more than 18 million Iraqis had one TV per 18 persons and one radio per 5 persons.

Iran

Iran started its radio system in 1940 and used it for decades as the key to political power, keeping it under government control. Television, however, began in 1958 under a five-year private ownership grant. It was later expropriated by the government as part of a growing national system, under the control of National Iranian Radio and Television (NIRTV). During the reign of the Shah in the 1970s, when the price of oil quadrupled on the world market, the flow of income resulted in the expansion and modernization of existing broadcasting facilities, including the importation of entertainment programs from the west. This modernization of Iran was one of the reasons that the conservative religious leaders sought—and succeeded in achieving—the overthrow of the government and the establishment of a religious state. The new leaders did not abolish broadcasting. In fact, the Ayatollah used the media very effectively in his seizure of power, sending thousands of deceptively labeled cassette tapes to be played in mosques to gather support for the removal of the Shah and to solidify his own power. Today all media is severely restricted and censored, with limited operating hours. All programming must extol the virtues of Islam and the government; English is prohibited and women may not appear on television. In 1995 Iran banned satellite dishes "to safeguard cultural boundaries of the country and of its families against destructive and indecent satellite programs" However, it had only one TV per 23 persons and one radio per 4.7 persons for its over 61 million people.

Lebanon

As noted above, Lebanon is unique in Middle East telecommunications. It has an extensive system of private TV stations, operating much as the commercial stations do in the United States. In 1993 there was one TV set per 3.4 persons and one radio per 1.3 persons among Lebanon's 3.5 million people. It was one of the early mideast countries to have broadcasting. The French colonial power established the first radio station in 1937, principally to counteract the radio propaganda from the Germans and Italians. When Lebanon gained independence in 1943, it put radio operations under the Lebanese Broadcasting Station in its Ministry of the Interior. However, the government didn't consider radio a priority in its national growth. When it was ready to authorize television, it took an approach entirely different from any of the other Arab states, either in Africa or Asia. It approved a request from a business group in Lebanon to supervise the construction and operation of a TV station that would be operated privately and financed by commercial advertising. The first private station, La Compagnie Libanaise De Television, was shortly joined by another private station, Tele-Orient. However, privatization did not go so far as to prevent the government from being certain that the stations did not threaten national security: the government controls the content of all news broadcasts, local, national, and international.

Jordan

Jordan also has a long history in broadcasting. One of the first radio stations in the Arab world was Jordanian Radio in Jerusalem, begun in 1936, broadcasting in three languages: Arabic, English, and Hebrew. During Israel's war for independence in 1948, Israeli forces captured the radio station, and Arab forces began operating an Arabic radio station in Ramullah. In 1956 the king of Jordan established a new radio station in Amman and, in 1959, a station in Jerusalem. The latter was taken by Israel in the 1967 war. Just as Israel established television following Jordan's introduction of TV, Jordan was earlier prodded into TV operations by the development of stations by Iraq and Saudi Arabia. In 1994, as a peace accord between Israel and the Palestine Liberation Organization was concluded and peace pacts with other Arab states seemed likely, plans were already underway to establish a Palestinian station in the self-rule areas of Israel, with Jordanian advice and assistance. In the early 1990s Jordan's 3.5 million people had one TV set per 12 persons and one radio per 4.5 persons.

Radia Alkhas, Director-General of Jordan Radio and TV Corporation, told us that he believes that Arab nations, as developing countries, should be more aggressive in distributing their programming (specifically news) internationally. This is, in fact, now happening. Many Arab countries are moving toward democratic, independent media, Alkhas says. For example, Jordan plans to work more closely with neighboring countries, including joint productions. Alkhas says the key to growth is more private sector production and transmission, and local stations. He predicts that there will be more local stations in the Arab world, in contrast to the current centrally controlled radio and television networks, with every city of 500,000 to 1 million people having its own station.

Jordan Deputy Director-General and Director of TV, Ibraham Shahzadeh, agrees. He told us that he believes there is a great need for regional cooperation, with joint production within the different Arab broadcasting unions, especially in the area of documentaries and cultural programs. Arab countries belong to three major regional broadcasting groups: Arab States Broadcasting Union, Islamic States Broadcasting Union, and Asia-Pacific Broadcasting Union (ABU). Emanating out of the ABU, Asia Vision provides satellite exchange programming. Many Arab nations are also members of the European Broadcasting Union (EBU). The Islamic Broadcasting Union has forty-five member countries, mainly concerned with joint radio religious programs in all languages of the region. Satellite technology has opened up the Arab countries to not only regional but worldwide exchange of information and ideas—and it is growing. Shahzadeh believes that the Arab states' broadcasting systems must move from being government owned to private ownership, to avoid political control and to serve their countries better. For example, at the present time the Jordanian director-general reports to the Minister of Information, as in most other Arab countries. Shahzadeh says: "The Jordan system should be state-owned, but with a Board of directors appointed by the King—somewhat like the BBC."

Jordan, which is more oriented to western concepts and practices of democracy than most, if not all, other Arab nations, has had some effect in introducing democratic concepts to other countries in the region through its satellite channel programming to other Arab states. In part, Jordan is able to do this because it doesn't have the problem of rebel Islamic movements as in other Arab states, because the Islamic movement in Jordan has been integrated into the country's political structure.

Syria

Syria's introduction to broadcasting was somewhat similar to Lebanon's—both countries were under French rule. Radio broadcasting began in 1946, a year after the country achieved independence. The economic condition of the country put radio on the back burner for years, and it wasn't until the 1960s that the Syrian government fully realized the political importance of radio and at that point began assiduously to develop it. Television began about the same time, following Syria's establishment with Egypt of the United Arab Republic (UAR) in 1960. However, the country's economy and the mideast wars prevented the government from committing the funds necessary for the media to grow. By the 1990s, however, Syria had a ratio of 1:17 for TV and 1:4.1 for radio sets in its population of almost 14 million.

Israel

Professor Ziad Rifai believes that Israel will play a major role in mideast broadcasting after the peace accord. He thinks it will provide exchange programs, although it won't be a member of the Arab group. He explains that 50 percent of Israelis came from Arab countries and would be interested in programming from Arab countries. Israel would also have an important technical influence among Arab states. Even now Jordan and Israel channels can be seen across their respective borders. In the south, Israel and Jordan can get Egyptian and Saudi Arabian channels, and Egypt can get channels from Jordan, Israel, and Saudi Arabia. Emmanuel Halperin, director of foreign news for Israel TV, has a different view. He told us that he agrees that program exchange is dependent in part on the segment of the population of any given country that came from another country, and that Israel has large percentages of people who came from other mideast nations. However, he believes that historical and attitudinal factors will preclude Israel from substantial exchange with individual Arab countries. He states that although an Israel-Egypt telecommunications accord dates back to their peace agreement of 1982, virtually no cooperation between the two countries has taken place. Halperin remembers that when Egypt signed the 1982 agreement, its representative said: "We have cooperation with 98 countries; you're the first one we have to sign an agreement with." "Result," says Halperin, "no cooperation." He believes that an overall Israel-Arab nations peace agreement would be needed, that agreements between Israel and individual Arab countries won't work.

Mordecai Kirschenbaum, Director-General of the Israel Broadcasting Authority told us, however, that he believes there will be cooperation among mideast countries, including Israel. He notes that in early 1994, for the first time, Egypt wanted to carry Israeli programs in Arabic. He notes, too, that there will be a new Palestinian station in Israel. He believes that Israeli broadcasting will reach a greater part of the world, as world systems change with expanded use of satellite and cable.

Israel's economy and western cultural ties have enabled its 5 million people to have TV and radio sets in most households, with a per person ratio of 1:4.1 for TV and 1:2.2 for radio.

ARABSAT

Professor Ziad Rifai notes that satellite use is growing in the mideast, but not necessarily for mideast populations. ARABSAT was launched in the early 1980s; at first it was underutilized, Ziad says; now countries are fighting for space. They beam signals to the region and even to Europe. About a dozen countries currently have space transmission, but not too many people in the region have satellite dishes. The key audience, Rifai says, is therefore expatriates in Europe and the United States. Some signals are being picked up and retransmitted on other countries' broadcast stations and on wireless cable systems, but not every country's signal is being picked up extensively. For example, Jordan's is not; Egypt's is. Professor Ziad estimates that it costs a country $1.2 million a year just for rental of space on ARABSAT. In addition, it has to produce and feed the signal, with programming costs alone running another $2-3 million a year, and all this is without commercial subsidization.

Jordan, Egypt, Tunisia, and Morocco have legalized satellite receiving dishes. The Gulf States have not legalized them, but have permitted people to have them. They are not legalized or allowed in a number of other countries, including Iraq, Syria, Iran, and Libya.

Forces of Control

The nature of governance in much of Africa sometimes creates telecommunications systems unrelated to the direct needs of the public the systems ostensibly serve. For example, as King Abdul-Aziz of Saudi Arabia got older, he found it more and more difficult to travel and maintain communication with leaders in other parts of the country. In 1949, therefore, he ordered a radio studio and transmitter built in Jeddah so that he might be heard throughout the land. This led to the development of a broadcasting system in the country.

Some countries use their telecommunications media as adjuncts of military dictatorships. For example, Sudan has state-controlled TV, which propagandizes with martial music and pictures of marching soldiers. In early 1994 the principal target audience was in the southern part of the country, where the government was fighting rebel separatists. Strong control of the media in almost all African countries by the government—that is, the royal family or the party in power—results in almost total censorship. The rationale, as in western countries, is, of course, national security. In Africa censorship is also justified as a means of precluding media material that might incite trouble and even violence between tribes. Not only does gov-

ernment ownership and control of the systems, in most cases, provide the means for content censorship, but in many African countries journalists are required to be licensed and, as such, are frequently considered employees of the government. They can therefore be summarily fired, or even fined or jailed, if they publish or air any materials that the government deems inappropriate or that have not been cleared with the government censor. As might be expected, this has led to highly effective self-censorship.

External forces, similarly unrelated to the needs of a country's people, also play a role in the development of telecommunications systems. The first TV station in Saudi Arabia was established in 1955 by the U.S. Air Force to entertain its personnel at the Dhahran air base. Within a decade Saudi Arabia had built its own television stations at Riyadh and Jeddah.

Even where there are relatively independent regions or affiliated states, the centralized governments maintain a strong hand in television and radio broadcasting. The United Arab Emirates (UAE), consisting of seven sheikdoms, is an example. Before affiliation into the UAE in 1972, some of the countries operated their own systems, evolving out of the broadcast stations that the British colonial power had initially introduced for its own use. The stations remained under the operation of the individual emirates: Abu Dhabi, Dubai, Sharjeh, and Ras Al Khaima. However, the central UAE government controls all operations and determines personnel and programming. In the early 1990s, cable had not yet made inroads into the UAE telecommunications mix.

Southern Africa

Southern Africa, like eastern Africa, by and large lacks the funds, technology, energy resources, and trained personnel to operate effective radio and TV systems. Nevertheless, every country has recognized the political value of radio and has established national networks, and many countries have also made TV operations a national endeavor. The systems usually operate with a main central station and transmitters and regional or local stations in strategic parts of a given country. The geographical and topographical makeup of many countries in Africa requires creative relaying of signals to cover a substantial part of their territory. For example, in 1990 Madagascar, an island off the coast of southeastern Africa, used one national station to feed thirty-six low-power relay transmitters. The Seychelles islands of Africa —about 100 of them make up the country—had two national stations distributed by nine scattered transmitters. Mauritius, another island off Africa's southeast coast, had one station and three relay transmitters.

South Africa

South Africa is the only country in the region not to fall under third-world status, and, as one might expect, has an intensive and extensive telecommunications system. In the early 1990s, it was operating twenty-three radio services broadcasting in seventeen different languages, and four TV services in seven languages. For its 42 million people there was one TV set per 11 persons and one radio per 3 persons. In 1991 it authorized its first independent, privately owned TV station. Until the advent of free elections and majority rule in 1994, the broadcasting system was run by the South African Broadcasting Company (SABC), which was under the control of the minority-led Afrikaaner government; in fact, for many years South African broadcasting was called by many the propaganda arm of the National Party. Whether the new government of Nelson Mandela will convert the country's broadcasting system to one of greater independence, serving all of the people, is not yet known. However the inital signs were promising. The South African Broadcasting Network was among the first of the federal orga-

nizations to be changed to meet the needs of the wider public. New community radio stations, operated by people previously disenfranchised from operation of or participation in broadcasting, were being devloped all over the country.

Bill Siemering, the recipient of a MacArthur "genius" grant, used it in the 1990s to work with European and South African broadcasters as they sought to develop new radio and television systems and models after the collapse of communism and apartheid. In our talks he expressed considerable optimism for the future of South Africa's audio medium. Says Siemering:

The nation has the potential to develop the most diverse and effective radio system in the world for these reasons: a) The high illiteracy rate gives radio dominance over all other media, b) The majority of the people have a rich oral tradition which is ideally suited to radio, c) The substantial development needs give high motivation for effective use of radio, d) The SABC, under new leadership, will undergo a complete transformation, e) Radio is the least expensive medium, thus is an appropriate technology for South Africa at this time, and f) Radio is personal, imaginative, and democratic in that it is accessible to all.

While Siemering is enthusiastic about prospects for the country's electronic media in the absence of the deposed white supremacist government, he has some reservations.

Out with the old, but what the new will bring is not entirely clear. Perhaps there is no other country where radio will undergo such a metamorphosis as in South Africa. The country has a once in a lifetime opportunity to create a new indigenous radio system. Of course, inexperience in station management can bring failure. Stations must be built according to solid architectural and engineering principles or the building will collapse. These new stations must be built upon firm foundations, and time will reveal if they have been.

Lesotho is an independent country located within South Africa's borders.

Nonetheless, it has its own station, Radio Lesotho, reaching the 20 percent who have sets out of its 2 million people. It also operates a TV system, but in the early 1990s less than 1 percent of the people had TV sets. Namibia, on South Africa's border, was until recently totally controlled by the latter. The Namibian Broadcasting Company operates radio and TV systems, with about 20 percent of its 1.6 million people having radios and about 2 percent having TV sets. Another South African border state recently breaking away from political domination by its neighbor is Zimbabwe. The Zimbabwe Broadcasting Corporation is an independent organization patterned somewhat after the BBC. It operates both TV and radio systems, with about 5 percent if its more than 11 million people owning radios and about 2 percent owning TVs. Botswana, also on South Africa's border, operates its own independent radio system, with virtually one radio for every person of its 1.3 million population. While it had not yet established its own TV system in the early 1990s, about 5 percent of its people amowned TV sets and picked up signals from neighboring countries. Mozambique, also on South Africa's border, has a radio system that is partly state-owned and partly privately owned, and commercially operated. However, only about 5 percent of the 15.5 million population own sets. In the early 1990s Mozambique began to experiment with TV broadcasting, which was growing at a very slow pace. Swaziland is yet another country that has been influenced by its border with South Africa. It operates radio and TV systems nationally, with almost 20 percent of its 1 million people having radios and about 2 percent having TV sets.

Central and Northern Africa

Many countries that have both radio and TV operations have economies too poor to permit people to buy enough sets for either service to make the broadcast systems effective. It is necessary for extended family groups and, in some cases, entire villages to gather around one radio or TV set in order for the media to reach even moderate numbers. Angola, with 9 million people, is one of these countries, with government-operated radio and TV transmitters covering the country, but with only one TV set per 200 people in the early 1990s and one radio per 22 people. The 3 million people of the Central African Republic have a better situation for their state-run radio system, with a 20 percent penetration. The country has no TV system, but as in some other countries, a small percentage of people have sets that pick up neighboring TV signals. Burundi, 6 million people, has state-operated radio and TV systems, but only 10 percent of the people own radios and very few have TV sets. Rwanda (8 million) is much the same, with identical radio ownership; as of the early 1990s, it had no plans for television.

Figures are generally similar for other countries in the area. Djibouti, a small nation of 400,000, has state-run radio and TV systems. Its radio penetration in the early 1990s was 6 percent, with TV ownership at 3 percent. Zambia's 9 million people have an independent radio system patterned after the BBC, the Zambia National Broadcasting Corporation, but a government-controlled television system, Television-Zambia. One in 44 persons owned tv sets and one in 14 radio sets in 1992.

Countries with about 10 percent radio ownership include the Congo (2.4 million population) and Zaire (40 Million). Nations with about 20 percent radio set ownership include Uganda (20 million population), Kenya (26 million), and Gabon (1.1 million). All have TV distribution but set ownership is minuscule, except for Gabon, which has about 3 percent penetration. Twenty percent of Ethiopia's 51 million people have radios, but less than 1 percent own TV sets to receive programs from the state-owned Ethiopian Broadcasting System. Both Malawi (10 million population) and Sudan (29 million) use their broadcasting systems effectively for government purposes, and in both countries about 25 percent of the people own radio sets. Perhaps because it is, at its closest point, 250 miles away from the mainland and other terrestrial broadcast signals, the island of Madagascar of 12.5 million people, technologically and economically weak, has had to develop effective radio and TV systems to cover the island. About 20 percent of the people have radios and 2 percent TV sets.

In a number of countries, continuing political fragmentation and sectional wars have made it impossible to develop national radio and TV systems. Somalia, with state-controlled radio and TV, is an example. Only 5 percent of its 7.3 million people have radios, and fewer than 1 percent TV sets.

It must be noted again that figures for per capita ownership of sets may be misleading. The poorer the country and the smaller the number of radio or TV sets, the greater the likelihood of large extended family groups, or groups of people in villages and rural areas, listening to or watching any given set. A 2 percent penetration of television, therefore, may translate in some villages into virtually 100 percent viewing, and in some countries into 25 percent or more availability; the same principle holds true for radio.

Many African countries that found radio to be of special value held off the development of television for many years because their limited finances suggested that a wiser course would be to strengthen their radio systems first. When they did develop television, principally beginning in the 1970s, it grew strongly, with a number of countries establishing national and/or regional and/or local distribution outlets, and eventually almost all countries began utilizing satellites for importation of programs. Nigeria, for example, established local TV stations, not only to facilitate distribution, but to produce programs representing the specific region's needs, with many of these programs then distributed through the national system. In the early 1990s Nigeria's population of 89 million had one TV set per 25 persons and one radio per 10 persons.

Despite its comparatively strong growth in the past two decades, TV has fallen far short of the miracles most African countries expected it would produce. They thought it would be the answer to providing universal education, to strengthening national cultures, to combating illiteracy, to vocational preparation, to agricultural advancement, and to preventive health measures, especially for the urban poor and the rurally isolated. To try to accomplish this, most countries used both satellite and terrestrial systems to achieve national coverage. Where there was no electricity—the case in most rural areas—the government provided and encouraged the use of solar-powered and manual-powered generators. Cote d'Ivoire (the Ivory Coast), for example, was almost immediately able to reach 70 percent of its 13.5 million population by providing government-owned TV sets for every village in the country for group-watching. (In the early 1990s its TV set ratio was 1:15 and radio set ratio 1:8.) Niger did the same for its 8 million people with UNESCO help and funds from France, distributing solar-powered TV sets to villages without electricity and organizing the viewing of selected programs by large groups. Its radio penetration was 5 percent. Camaroon set up a system that covered 90 percent of its 13 million population, relying on British and French help to do so. Camaroon's radio penetration was 20 percent.

While technologically advantageous, reliance on the west proved ultimately unsound. Without a tradition of electronic telecommunications, most African countries depended on the west for guidance and training. They copied western approaches to organization, production, and programming, which didn't necessarily work in an African country. They didn't have the monetary resources necessary to develop their own applicable technology, production approaches, and knowledgeable personnel. They had no preparation in setting up an effective infrastructure suited to their individual needs. Ultimately, all of these factors, plus the continued poverty of most of the potential audiences, resulted in the failure and discontinuance or the languishing of TV systems in some countries, with the continued growth and strengthening of TV in only a few. To too great an extent, TV became available only to the elite.

Most of the countries in Africa with TV systems have satellite earth-receiving stations, distributing programs that they deem suitable from other African countries and, in some cases, from the U.S. program-dominated STAR and SKY satellites. Satellite use created ironic shifts in the structure of individual countries' systems. At first, most systems were highly centralized. Gradually, most countries began to decentralize and establish regional and local stations. With the increasing use of satellites, there is a trend back to centralization for national selection and distribution of imported programs.

So-called "pirate" radio stations appear and disappear frequently in many countries in Africa (with the spread of privatization, the market for commercial pirate stations has declined worldwide). Unlike the pirate stations of Europe, which seek to capitalize economically on the public's demand for popular entertainment by presenting highly commercialized programming, such stations in Africa are principally politically oriented, clandestine stations. They operate usually on very low power to reach the people in a politically volatile region and move frequently as government forces try to track them down. The dictatorial control in most African nations, and the revolutionary efforts in many, give rise to these stations.

One of the problems in program-production cooperation and exchange in Africa is the incompatibility of systems. In large part due to their former colonial occupiers or recent economic and political allies, about half the countries in Africa use PAL and half SECAM.

ASIA

The mideast countries of Asia already have been dealt with in this chapter. Most of the other countries in Asia differ markedly in some ways from the telecommunications systems in the mideast, and in some ways are surprisingly similar.

There are the growing or already economically advanced nations, such as Japan, South Korea, China, Taiwan, Singapore, and Hong Kong, all industrialized with extensive telecommunications systems, although China has only recently taken its great economic leap forward and much of its population is still poor and rural. Most of the other nations in the area are third world in economy, growth, and telecommunications. As in other regions, some countries have decentralized combinations of public and private stations, while others principally serve the government, with frequent political upheavals preventing the free growth of telecommunications. The so-called first-world and second-world nations have not only broadcasting, but satellites and, in some, growing cable systems. The third-world nations have radio, but until the 1990s only a few had viable TV systems with substantial penetration in terms of area or population. During the first half of the 1990s, television homes in Asia increased by more than 70 percent. In 1993 the ASIASAT satellite was launched, with a footprint covering most of Asia and part of the Pacific, and with the potential for future DBS service. Cable also began to grow, with India, for example, using thousands of cable systems to provide television to small communities. Other countries increasingly began to solve the problem of servicing local areas in the same way.

Japan

Japan is not only the leading telecommunications entity in Asia but, with regard to technology, in much of the world.

For example, it has taken the lead in high-definition, satellite-delivered television. Some observers say that Japan's system is the most efficiently organized one in the world. In 1977 WARC (the World Administrative Radio Conference) allotted eight frequencies to Japan for satellite use, which ultimately added some sixty additional channels to the extensive system of state-operated and privately owned stations. The Japan Broadcasting Company (NHK) is the country's national public network (Figure 2-2). Individual provinces have public stations as well. Five private TV networks, with numbers of individual station affiliates, covered the country's five islands in the early 1990s. Japan, with a population of 125 million, is among the world leaders in the penetration of radios (one per 1.2 persons), TV sets (one per 1.6 persons), and VCRs (almost three-fourths of the people). This translates into TV, radio, and VCR equipment in virtually every household. As of 1992, cable had not yet made any appreciable inroads in Japan.

South Korea

South Korea's area is small and was easily covered in the early 1990s by one government and one private TV network and by public and private radio networks. In the early 1990s, the country was also adding cable and DBS to its telecommunications distribution systems. The thriving economy of this country of 44 million people accounts for a high penetration of sets: one TV set for every 5 people, and one radio for every person. Like Japan, South Korea uses NTSC.

China

China has the most extensive national system of broadcasting in Asia, using satellite transmission to cover its vast territory, and at the same time it has possibly more local stations than all other countries combined. Under the Ministry of Information are China People's Broadcasting Station (CPBS), the

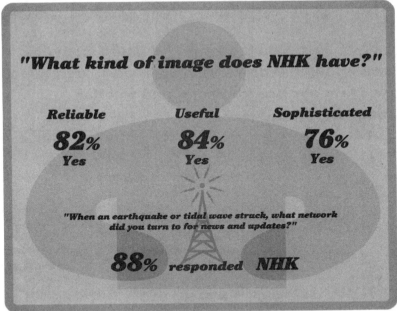

We surveyed*our viewers and listeners:

"What kind of image does NHK have?"

Reliable
82%
Yes

Useful
84%
Yes

Sophisticated
76%
Yes

"When an earthquake or tidal wave struck, what network did you turn to for news and updates?"

88% responded NHK

* September 1989

FIGURE 2-2
Japan's NHK conducts public-image surveys of itself. Courtesy NHK.

national radio system; China Central Television (CCTV), the national TV system; Radio Beijing, which serves the capital city; and the Central Broadcasting and Television University (CBTVU), which provides formal educational opportunities for millions who otherwise would not be able to obtain any higher education. In addition, there are many regional and local TV stations, all under the supervision of CCTV. While ostensibly adhering to the programming dictates of Beijing, many local stations, principally oriented to local education and news, produce programs that more specifically serve the interests and needs of their viewers, even if they are not always consistent with what the central authority prescribes. Because the country is so large, there is usually no problem or penalty attached to such independence, unless the content is so controversial that complaints about it reach Beijing. Radio is probably the most influential medium in China. In addition to the national networks, the over 100 provincial networks, the myriad of local stations, and the international service, almost every work unit—the organization to which every person is assigned when entering the work world—has its own radio station or stations. These wired radio networks were developed in 1950. There are literally hundreds of thousands of such intercom-like radio stations in China. They are connected to individual living quarters, offices, factories, and open areas, and invariably serve as early morning alarm clocks, playing martial music followed by a political lecture. They also play music to accompany the millions of Chinese who gather on street corners and other places early in the morning before going to work to practice tai chi. Because of group use of individual sets in many instances, the ratio of receivers in the population of 1.2 billion in the early 1990s was one TV per 8 persons and one radio per 9 persons. With its economic book and market economy moves, ownership of TV and radio sets increased markedly in the 1990s.

Before the 1949 peoples' revolution and the establishment of the Communist government, all stations—only radio existed then—were run for the political or economic purposes of their owners: political parties, foreign business interests, and wealthy individuals. The post-1949 era established four principles of media responsibility: to keep the people informed of news and current affairs, to disseminate and promote the policies of the government and the Communist Party, to orient the news to support the purposes of the government, and to motivate patriotic work habits in the interests of the country as a whole. In fact, when TV was established in 1958, its official charge was to educate the people, the Army, and the Communist Party to "build socialist spiritual and material civilizations."

China is rapidly expanding on the international scene. It has begun to import some programs for use in major cities where there are foreign tourists. The following table shows Chinese cites servide by satellite TV service:

City	STAR TV (,000) Homes	% of TV Homes
Chongqing	3,172	85
Chengdu	1,662	72
Zhenzghou	1,017	72
Kunming	831	95
Guangzhou	723	49
Changsha	536	39
Nanning	510	80
Xi'an	405	26
Changchun	396	25
Guiyang	342	88
Shijiazhuang	329	46
Jinan	310	23
Beijing	283	9
Qingdao	209	12
Hefei	199	21
Nanjing	143	11
Tianjin	131	6
Harbin	123	9
Shanghai	103	3
Nanchang	80	9
Shenyang	57	4
Xiamen	40	14
Dalian	35	3
Taiyuan	30	5
Ningbo	19	1
Hangzhou	7	1

Source: Frank Small & Associates; courtesy STAR TV.

For example, hotels can pick up CNN. (One of the authors of this book found in 1992 that whenever he tuned in to CNN in China, the picture and/or the sound was unclear; this raises the question of whether stories were screened before being made available, clearly and fully, to the viewers.)

Chen Guhua, Head of the Program Department of China Central Television (CCTV), told us that CCTV planned to develop further program exchange with western countries as well as expanding the exchange with other countries in Asia. Zhang Jianxin, anchor for the External Service, informed us of CCTV's plans to provide suitable programming for satellite distribution in many parts of the world as well as in Asia. As in Africa, systems are not necessarily compatible among Asian countries; China, for example, uses PAL.

Hong Kong has both private and government-supported broadcast stations. The latter, as might be expected given the long-time British control of the territory, is modeled after the BBC. As an international city, Hong Kong receives a substantial amount of satellite programming. Its excellent economy has resulted in high saturation of radio (one for every 2 people) and TV (a set for every 3 persons), with more than half the homes having a VCR. Programming is eclectic, representing many points of view. In 1997 Hong Kong officially returns to Chinese governance, which may well change the complexion of its telecommunications systems.

Other countries in the east and southeast regions of Asia are usually considered third-world countries, excluding Singapore and Taiwan, whose economies place them closer to Hong Kong than to most of their other neighbors. Taiwan, with 21 million inhabitants, has operated in the past with private networks that were closely monitored by the government. In the early 1990s, hints of political reform in Taiwan suggested that there may be changes that will allow the communications systems to expand, with more heterogenous programming. The expected introduction of cable, coupled with satellite reception, will open up much of the rest of the world to the Taiwanese. In 1995 plans for expansion and liberalization of government control were underway. Taiwan's set-penetration ratio in 1992 was 1:3.2 for TV and 1:1.5 for radio.

Indonesia has also experienced expansion in telecommunications as its economy has grown. In addition to the government-operated system, Indonesia has a pay-TV network that features programs from the United States, Europe, Japan, and Australia. Satellite dishes (there is one government-controlled satellite, and a number of people have DBS reception) and VCRs are proliferating. Thailand's system has grown during the periods of political stability—stability that has shaped the country to a greater degree than has been possible for many of its neighbors. It has developed one government-operated and four private TV networks, with appreciable help from Japan. In the early 1990s Thailand was planning to bring in cable systems.

Continuing civil strife in Cambodia has prevented the development of comprehensive radio and TV networks, leaving a government-operated radio system and a highly limited TV operation. With indications of possible peace in the country by the mid-1990s, telecommunications may be given priority. In the early 1990s Cambodia's 7 million people had one radio set per 10 persons but only one TV per 100 persons; Thailand's 58 million population had one radio per 5 persons and one TV per 15; and for Indonesia's almost 200 million people, it was one radio per 8 persons and one TV per 20 persons. Cambodia, Thailand, and Indonesia all use PAL.

Separated from virtually all of its neighbors more by politics than geography, North Korea (22 million people) has a tightly controlled radio and TV system, with one radio for every 4 persons, but with TV ownership only at about 2 percent. While South Korea uses NTSC, North Korea uses PAL. For some years politically isolated like North Korea, Vietnam in the early 1990s began to establish trade and travel relations with other countries, particularly the United States and other western nations. It is expected that its TV system, consisting of one government-operated network with limited range and air time, with only about 4 percent of its 69 million people owning sets, will begin nationwide expansion. At the present time both NTSC and SECAM sets are in use, to pick up signals of neighboring NTSC countries as well as the Vietnamese SECAM system. Radio, with one in ten ownership in the early 1990s, is also expected to expand.

While relatively few people in this region of Asia presently receive satellite programming—programs that come principally from U.S. and/or Australian satellites—international programming use is expected to grow rapidly as more and more political and trade barriers are eased.

The 1990s are considered the dawn of the satellite era in much of Asia, with the prospect of satellite distribution playing an important role in most countries by the end of the decade. Satellite organizations and cooperative projects were underway in the early 1990s.

India

Not surprisingly, the Indian broadcasting system is patterned after that of the British; it also retains the rigid control of the media that the British established as a colonial power. Radio goes back to amateur experimentation in 1924 and the first stations in 1927. The Indian government, when still under British rule, officially began to operate the radio stations in 1934. With independence in 1949, control was turned over to the new Indian-run government, but the British practice of tight control of radio for political purposes was retained. UNESCO started the first television broadcasting in 1959, and a second station was added in 1972. In the early 1970s India was able to use a U.S. experimental satellite to provide educational materials and news to selected isolated rural villages, as well as to some cities. The impact of television was carefully considered by the government in the 1970s, and any TV distribution and programming was carefully evaluated, sometimes for political purposes. In 1970 an Indian official explained to one of this book's authors the government's fear of free and indiscriminate use of television, especially to the huge masses of impoverished people: "How are you going to keep them down on the farm after they've seen TV?" Little has changed in the intervening years. By the 1990s the country's 886 million people had a set per person ratio of 1:44 for television and 1:16 for radio.

Pakistan

Pakistan, having been part of India prior to 1947, follows the same pattern of media control and operation. Pakistani radio was begun in 1925 by King Amanullah Khan as a way to give his country outreach and prestige on the international scene. The major cities' radio stations were in place prior to 1947. Pakistan developed an extensive system of regional stations. Television was developed in the mid-1960s, at first under private auspices, but eventually under government control. Under changing regimes and political systems, broadcasting has remained under the government, and is currently operated by the Department of Culture and Information (see Appendix, p. 52). In the early 1990s one out of 60 of the country's 121 million people had a TV set and one out of 12 had a radio set.

Bangladesh

Bangladesh's broadcast history is similar, having been a part of India before 1947 and a part of Pakistan before it was granted independence in 1971. It operates a system of radio and television stations, although, as in India and Pakistan, the economic conditions of the vast numbers of its 119 million people preclude much ownership of TV sets. In 1993 only one in 24 had a radio, and fewer than 1 percent had a television set.

OCEANIA

Australia and New Zealand

This area of the South Pacific, which comprises Australia, New Zealand, and thousands of inhabited islands (in this book we will refer to the area as Oceania), is too often ignored in western geography books and in the media. Yet Australia has a significant impact on Asian media, and New Zealand at one time, before it privatized much of its media, served as a model for fairness, the objective presentation of material, and equal opportunity for different points of view.

Both Australia's and New Zealand's public or government systems are modeled after the BBC: government funded but independently run through the Australian Broadcasting Commission (ABC) and the New Zealand Broadcasting Company (Figure 2-3). Both countries' national public systems compete with privately owned stations. Privatization in both countries is growing at a rapid pace in the 1990s. Key broadcasting properties in both countries, for example, are owned by media baron Rupert Murdoch. (Murdoch was originally an Australian citizen, but in order to meet FCC requirements for station ownership in the United States, he became a U.S. citizen.)

Australia's satellites serve not only that country, but New Zealand and the myriad Pacific island nations as well. In 1993 Australia's ABC launched a satellite service to Asia. Australia's many isolated areas and vast expanses, such as the Outback, are reached by satellite. Australia's population of 18 million is virtually saturated with TV and radio sets: one TV per 2 people, and the number of radios exceeds the total population. In many isolated areas there is no electricity for reception, and generators are used. In a number of Australian films distributed globally, such as *My Brilliant Career,* one or more characters in an isolated area are seen attending Australia's "School of the Air" via two-way radio. In a visit to the school's headquarters in Alice Springs—located in the exact center of the country—we observed and heard classes in session, a model for long-distance education. The advent of DBS added a TV dimension to the School of the Air, which previously had loaned receivers and tapes, and now can conduct visual lessons with two-way audio.

New Zealand, a small country in area with fewer than 4 million people, has radio and TV systems that completely cover both its north and south islands. It, too, has more radios than people, and one TV set per three people—that is, about one per family. Its satellite service, as noted earlier, comes through the Australian satellite system. (See Figure 2-4.)

The Islands

Most of the hundreds of islands in the south Pacific have their own radio systems and some have cable-distributed TV. Most receive their radio and television signals principally from Australia, some signals from New Zealand, and the

FIGURE 2-3
New Zealand's Te Reo Tataki fact card. Courtesy TVNZ.

TELEVISION NEW ZEALAND
TE REO TATAKI

- Television New Zealand operates two complementary channels – TELEVISION ONE and CHANNEL 2 – and is New Zealand's leading national broadcaster, reaching almost 100% of the country.
- TELEVISION ONE is transmitted from the Avalon Television Centre in Wellington, and CHANNEL 2 from the Auckland Television Centre.
- TVNZ comprises many divisions ranging from programming, production and international sales to telecommunications and transmission services, sales and marketing, news and current affairs and Teletext.
- Over 12,000 hours of television are broadcast each year on the two channels. Local programming accounts for approximately 30% of total programming, and in 1991 amounted to 3,255 hours.
- Internationally, TVNZ's natural history programmes and family dramas have won many awards, and are screened in over 40 countries.
- TVNZ employs over 1000 staff around the country, in the main centres of Auckland, Hamilton, Wellington, Christchurch and Dunedin.
- The Auckland Television Centre is the company's headquarters, housing state-of-the-art production and transmission facilities. Address: 100 Victoria Street, Auckland. P.O. Box 3819, Auckland Tel: (09) 377 0630, Fax: (09) 375 0979

TELEVISION
NEW ZEALAND

FIGURE 2-4
New Zealand's educational TV service.
Courtesy ETV.

◆ WHAT IS e^{TV}?

Education Television (e^{TV}) is an initiative of TVNZ. It broadcasts and produces high-quality educational programmes. e^{TV} aims to help meet the growing needs of New Zealanders for life-long learning resources, through the provision of the best of international programming, as well as local productions.

e^{TV}, in association with major institutions, has introduced the concept of Television Learning courses for credit. In 1994, e^{TV} offered *French, Spanish, Statistics, Child Development* and *Marketing* through Auckland Institute of Technology and *World Religions* through Massey University. We see this as a major source of growth as the demand for learning resources increases in our society.

e^{TV} broadcasts daily, currently offering 12.5 hours of programming per week. There are general themes underlying daily schedules: Monday:*Asia*; Tuesday: *Environment and Social Issues*; Wednesday: *Women and Relationships*; Thursday:*Science and Technology*; Friday: *Culture and Humanities*; Saturday: *Business*. On Sundays, *Asia Dynamic* focuses on the life and culture of New Zealanders of Asian descent.

◆ IS e^{TV} FOR YOU?

e^{TV} is for everyone interested in expanding their range of interests or improving their qualifications. The concept of continuous learning is an international trend that is being fostered by e^{TV}. e^{TV} is *not* schools broadcasting (though many programmes are being utilised in schools throughout the country).

If you are interested in programmes that stimulate and involve, e^{TV} presents quality options. If you are interested in 'up-skilling', an increasing range of courses leading to qualifications awaits you. In addition to broadcasts, e^{TV} is becoming a provider of learning materials in video, electronic and print forms.

◆ HOW CAN I FIND OUT MORE?

For further information, contact:
e^{TV}
TVNZ
PO Box 3819
Auckland
New Zealand
Tel. 09-375 0736
Fax. 09-375 0860

remainder from some of the larger islands that have broadcast systems, such as the United States' Western Samoa. Because of the vast distances in the island chain of any given country, and the general low economic resources of most of the islands, radio is virtually universal, while television usage is generally sparse. In French Polynesia, for example, radio reaches almost 100 percent of the inhabitants, but television penetration is only about 30 percent. For many islands, TV is nonexistent, except for those few wealthy enough to have a receive-satellite dish. In some others, TV began to grow in the early 1990s. For example, although Papua New Guinea, like several other islands, had many isolated people with no electric power and only three sets per thousand inhabitants, it had three TV stations in operation serving its 10,000 receivers. On the other hand, its radio system of fourteen national and twenty-one regional stations broadcasts in English and some thirty other languages to a ratio of one set per 20 persons. As is done in Micronesia, many islands tape Australian and United Nations programs and then edit them to delete materials their governments do not want to reach the people.

The lack of television on many islands such as Tonga and Papua New Guinea has spawned a vast network of movie houses, some of them portable cinemas that make a circuit of a given island. Those islands that have the economic resources and the technology for television reception rely heavily on imported films for the systems that have stations, and principally on satellite reception for those without a terrestrial distribution system. In addition to satellite-distributed programs from Australia and some from New Zealand, programming comes from Japan, the United States, and the United Kingdom. In the 1990s, with governments in Australia and New Zealand becoming more conservative, and New Zealand selling many of its utilities and resources to private companies, the expansion of private control and/or influence has changed public-interest programming and public-service broadcasting to commercial advertising properties and programming reflecting the private owners' views.

Appendix to Chapter 2

PAKISTAN TELEVISION TODAY

Sajid M. Qaisrani

Controller Public Relations, Pakistan Television Corporation

Pakistan Television Corporation Limited (PTV) is a Public Limited Company. All its shares are held by the Government of Pakistan (GOP). The decision to establish a general purpose television service with the participation of private capital and under the general supervision of the Government of Pakistan was taken in October 1963. Subsequently the GOP signed an agreement with the Nippon Electric Company of Japan, allowing the company to operate two pilot stations in Pakistan. The first of these stations went on the air in Lahore (Capitol of Punjab) on November 26, 1964.

Upon the completion of the experimental Phase in 1965, a private company called Television Promoters Limited was set up. It was converted into a public limited company in 1967. Television Centers were established in Karachi and Rawalpindi/Islamabad in 1967 and in Peshawar and Quetta in 1974.

PTV consists of ten divisions (News, Current Affairs, Programs, Sports, International Relations, Engineering, Finance, Administration and Personnel, PTV Academy, and Educational Television), each headed by a full-time Director.

There are presently six Television Centers, each headed by a General Manager. They are located in Lahore, Karachi, Islamabad, Peshawar, Quetta, and ETV Center Islamabad (PTV-2).

The majority of programs are relayed through a national network using terrestrial microwave facilities provided by Pakistan Telecommunication Corporation. The network links PTV's five Centers and thirty-two high-powered rebroadcast stations.

PTV reaches eighty-six percent of the population and covers about thirty-eight percent of the country's land mass. We broadcast programs in nine languages, including Urdu, English, Punjabi, Sindhi, Seraiki, Pushto, Hindko, Balochi, and Brahvi. Newscasts are also broadcast in these languages.

PTV is on the air eleven hours and forty minutes daily. We originate the majority of our programs. That is, only fifteen percent are imported. Our broadcast system is PAL B/G, and about sixty-three percent of our income is generated through commercials. Twenty-two percent comes from license fees and eight percent is generated by the sale of programs. Income from miscellaneous sources constitutes seven percent.

PTV inaugurated its Educational Television service in November 1992. It operates six hours daily, and its mode of telecast is ASIASAT. Sixteen Rebroadcast Centers in the country transmit programs relayed via satellite, thus covering twenty-four percent of the area and fifty-six percent of the population.

3 Control and Regulation of World Systems

Depending on the political orientation and aims of a given government, telecommunications systems vary from total state control to total privatization of ownership. As noted earlier, most of the countries in the world today either have full or quasi-dictatorships or are fragile emerging democracies; in both of these situations the political leaders almost always maintain strict supervision of the media. This is especially true of the electronic media, inasmuch as most of these countries have high rates of illiteracy, making radio and, where available, television the principal means of mass communication. There are three principal types of telecommunications systems in the world: state-controlled monopoly, public authority, and private ownership. As already seen, government-owned and -operated systems are endemic to monarchical and dictatorial societies, even those ostensibly democratically based. The British system and those of its Commonwealth nations illustrate the government-financed, public authority system. The United States exemplifies the system of private ownership, financed by advertising.

In both the state monopoly and public authority systems there are frequently mixes of private ownership, some on a par with the dominant system, some more or less competitive, depending on the degree of government need for control. Even in some systems where there is no private ownership, the concept of advertising has been adapted for purposes of financing. Some countries have a mixture of all three basic approaches. Government control or regulation also varies, with state systems usually under the direction of a Ministry of Information or a Ministry of Communications, whose purpose is to see to it that the media are used to inform and educate according to the tenets of the party, individual, or, in the case of a monarchy, the family in power.

Public authority systems are usually run by an independent committee or organization, sometimes on a national basis, as with the BBC, sometimes as regional systems, as in Germany.

Principal funding usually comes from fees levied on set owners and administered by the government. In some instances, for example in the United States, the public authority system is funded by general tax monies.

Public authority systems are usually noncommercial and nonprofit and are oriented to educational and cultural programming, their entertainment programs ostensibly reflecting high artistic standards. Privately owned systems are unabashedly commercial, inasmuch as they rely, with few exceptions, on advertising to keep them in operation. Their need to attract advertisers requires programming that attracts the highest possible numbers of listeners and viewers. Many private system owners have determined that programming that appeals to the lowest common denominator (LCD) automatically attracts the largest audiences, and therefore those systems are oriented to that level of popular entertainment. Private systems are usually regulated by a government body, with limited powers that are principally concerned with technical matters. In many cases, these agencies are insulated from overt day-to-day control by the party in power. In countries where private ownership has been permitted as an adjunct to state monopolies, regulation is usually much stricter.

NORTH AMERICA

United States

Systems and controls vary greatly in North America, reflecting the kinds of constitutional guarantees discussed earlier. By and large, because of the influence of the United States, most North American countries have, either entirely or in part, private systems with minimal governmental control or regulation. Canada's influence has helped lead to the creation of modified public authority systems in some countries in the region. In those countries with dictatorships or military rule—regardless of whether or not the country is a self-styled democracy—the government maintains strong control over state systems and any private stations it has authorized. The reverse is true in some countries where U.S. influence has prompted U.S. entrepreneurs to set up private systems, and to help establish individuals or organizations that in effect regulate the systems with the government's blessing. The United States has extended the structure and operation of commercial communications well beyond its own borders.

The basis for regulation of telecommunications in the United States is the "scarcity" principle. That is, there is a limited amount of frequency space, and in order to prevent complete ownership or domination of the airwaves by the few wealthiest, there must be some system of allocation and control. A second reason for regulation, under the scarcity theory, is the probability of stations canceling each other out through interference unless specific frequencies are assigned to specific licensees, and rules governing signal interference are strictly enforced. The early days of radio broadcasting in the United States, during which the Department of Commerce licensed but had no authority to regulate radio, resulted in chaos on the air and forced Congress to enact the Radio Act of 1927, which established the Federal Radio Commission to clean up the airwaves. The first volume in Erik Barnouw's landmark trilogy on the history of U.S. broadcasting is aptly titled "A Tower in Babel." Later, Congress enacted the Communications Act of 1934, which established the Federal Communications Commission (FCC) as the regulatory body for all telecommunications. The act has been frequently amended since, whenever Congress determines that new technologies or new social, political, or economic factors require a change in the regulatory law. The FCC itself constantly issues new (and amends old) rules and regulations; these regulations have the force of law, subject only to judicial appeals or further congressional amendation.

The FCC regulates both interstate and foreign commerce that uses the airwaves and/or wire for communications. Freedom of speech and the press is guaranteed not only in the U.S. Constitution's First Amendment, but in the Communications Act of 1934, Section 326 of which states that

Nothing in this Act shall be understood or construed to give the Federal Communications Commission the power of censorship over the radio communications or signals transmitted by any radio station, and no regulation or condition shall be promulgated or fixed by the Commission which shall interfere with the right of free speech by means of radio communication.

The act, however, also mandates that all licensees must operate in the "public interest, convenience, and necessity." This clause has given the FCC power to regulate virtually all aspects of telecommunications, with the exception of pre-censoring, or exercising prior restraint, of specific materials. But even here, the Communications Act authorizes the FCC to take action against any station that broadcasts obscene, indecent, or profane material—something which the FCC has done, after the fact.

The licensing power of the FCC is significant. While the FCC cannot regulate networks, it does regulate individual licensees. By establishing requirements for licensee behavior, it can directly affect the licensees' relationships to networks and therefore indirectly regulate networks insofar as they relate to their owned-and-operated (O&O) and affiliate stations. All telecommunications operations must obtain a license to go on the air, whether they are broadcast stations, microwave stations, or private short-range communications used for business, industry, public safety, or health purposes. In fact, by far the greatest number of licensees are those using the so-called private radio frequencies. In addition, the FCC regulates common carriers, such as telephone companies and satellite operators, and through the Cable Act of 1992 regained some of its earlier regulatory powers over cable. As of mid-1995 the following number of stations or systems were in operation in the United States: 1,170 commercial TV stations; 365 noncommercial, or public, TV stations; 10,055 commercial radio stations; 1,750 noncommercial radio stations; 1,625 low-power television stations; and 11,300 cable systems with 66.5 percent national penetration.

Using the scarcity principle as a base, the FCC has from time to time, depending on the philosophy of the political party in power, taken strong action against monopolies. There are rules governing multiple ownership (that is, ownership by one entity of a number of stations), duopoly (owning two or more stations in one community), and cross-ownership (owning a daily English-language newspaper and a broadcast station in the same community). Further, licensees must renew their licenses at periodic intervals (for example, seven years for radio stations and five years for television stations). In mid-1995, under the deregulatory philosophy of a Republican Congress, many of these rules were in the process of being loosened or abolished.

Part of the anti-monopoly rationale is the fact that all domestic systems that can be used by the public at large are privately owned. The federal government does not own or operate any domestic system that can be used by the general public; it does operate some private stations for government use only, in such departments as Defense, Interior, and Commerce, and the Department of Interior operates some noncommercial radio stations to serve isolated communities in Alaska. The only exception is the ownership of some public nonprofit, noncommercial radio and TV stations by some states and municipalities. At one time in its recent history, as a result of FCC appointments made by President John F. Kennedy and a subsequent national atmosphere of consumer-oriented civil rights and civil liberties, the FCC instituted various requirements, such as the Fairness Doctrine and the Ascertainment of Community Needs, which guaranteed the opportunity for the public to hear all sides of an issue and mandated that stations devote time to important public affairs and events in the station's community. The FCC also set limits on commercial advertising. However, strong deregulation in the 1980s under President Ronald Reagan eliminated most of the regulations designed to protect and serve the public, in favor of a marketplace system that returned much of the power to the private owners. One trend continuing into the 1990s is the relaxation of antimonopoly rules governing multiple ownership nationally and in individual markets. In the 1990s the FCC faced special regulatory challenges because of the increasing growth of new technologies that affect the operations of existing media and in themselves require new FCC rules.

Canada

Canada, as noted earlier, operates under the general system used by British Commonwealth countries. Its governmental jurisdiction is the Canada Radio-Television and Telecommunications Commission (CRTC). Canada has attempted to maintain a BBC-type public authority through its Canadian Broadcasting Company. However, in searching for more diverse programming, it also adopted the United Kingdom's introduction of private stations. Principally, during the 1980s, it promoted the growth of cable, and the majority of Canadian TV viewers are now served by cable.

The Canadian Broadcasting Act of 1968, which reorganized the country's broadcasting system, attempted to strengthen Canada's cultural unity and, concomitantly, its political and economic bases. To some degree Canada hoped that a unified, strong telecommunications system could help prevent the growing impetus toward withdrawal by the French-speaking part of Canada from the rest of the country. It also hoped to forestall the gradual incursion of U.S. ideas and ideals into the country, principally in the populous border areas, via easy reception of U.S. radio and TV signals. Indeed, it was the persuasive influence of U.S. television, which offered LCD entertainment choices that the CBC did not, that prompted some degree of private ownership in Canada. In effect, the entire country did what individual stations and networks do when the competition is growing too strong: counterprogram with similar materials that will hold one's audience. Even so, the programming of private stations is mainly imported from the United States, and cable offerings are mainly those of U.S. cable networks. A further U.S. spur to the development of advertiser-supported private stations in Canada was the advertising of Canadian companies on U.S. stations whose signals were preferred north of the border, thus sending funds out of the country. While, as noted later in relation to Central and South American countries, U.S. cultural imperialism has had a great effect on poor countries with limited communications resources, it is so strong that it can affect a relatively affluent and politically independent neighbor, as well.

In mid-1994 Canada had two state-run public TV networks, one broadcasting in English and one in French, and twenty-one private networks. Four pay-TV networks and 136 regional and local stations also served Canada. Canada's cable penetration was over 70 percent and its sixteen satellite-to-cable services included highly popular music-TV.

Central America

Government control and cultural imperialism combine in most Central American countries to limit the freedom of the media. To some degree this is because of the lack of clearly defined communications laws in general, and either tight, arbitrary government control or ad hoc strict reactive control in many countries in particular. Additionally, media freedom is limited by the political and economic influence of foreign capital, principally from the United States, and from foreign companies, also mainly from the United States, that support many broadcasting systems through advertising. Because of the faltering economies in many Latin American countries, their communications systems, both hardware and software, are dependent on foreign investors. Thus, there is further direct outside influence on structure, operations, and programming. A few countries whose economies are better than those of their neighbors, such as Mexico and Venezuela, are able to operate more independently of foreign influence. Privately owned commercial stations dominate broadcasting in most Central American countries, generally following the pattern of U.S. commercial stations. As in the United States, commercial advertising is the economic base for Central American operations. About two-thirds of all stations, both private and government, are commercial. In fact, it is estimated that more than 25 percent of total air time in Central America is devoted to commercials. (In the United States commercials consume over 30 percent.) The remaining stations are owned and operated principally by the government or by public groups. Privatization in telecommunications is growing through the increase in cable and satellite distribution, both of which are operated by private entrepreneurs. Because both of these services require additional receiving equipment and/or connection and subscription fees, the less affluent viewers are losing access to a growing number of television outlets, as happened in the United States with the growth of cable and its subscription fees.

While the concentration of television ownership in Central America is usually limited to a few powerful companies, sometimes to only one entity, radio ownership in many countries is more diverse. Obviously, the lower cost of radio operation opens it up to a broader economic spectrum. However, most radio stations are affiliated with a national network or organization, partly because it allows them to obtain programming at lower cost and partly because various governments encourage such affiliation as a means of facilitating government supervision and overt or tacit control over the media.

Programming is principally popular entertainment, concentrating on soap operas (a number of individual countries produce their own highly popular soaps), sports, and music (MTV spread at least as quickly in Central America as it did in the United States after its introduction in the early 1980s). About one-third of Central American programming comes from foreign sources, mostly from U.S. government funding—that is, taxpayers' money—which is used in many

countries to maintain the basic nation-wide distribution operations. About four-fifths of the 80 percent of Central American countries that have television use the U.S. NTSC standard. With the exception of those TV stations owned by governments, programming is usually independent of direct government supervision. However, as indicated earlier, the dependence of stations on government licensing, regulation, and in some cases, funding, plus the dangers inherent in programming materials that might be considered inappropriate by any given military regime, results in program content that is rarely, if ever, inimical to that which the government favors. A notable example of direct control is Colombia, which maintains strong supervision over all media content, even on the private stations licensed under its federal office, Intravision.

Because of the size of the countries and the lack of funds for state-of-the-art technology, most stations in Central America have low-powered transmitters and concentrate on local coverage. While the United States, Canada, and Mexico may be designated as large TV markets, only Costa Rica, of the other North American countries, has nationwide TV coverage. Guatemala, Honduras, and Nicaragua may be said to have medium markets, while the others, including El Salvador and Panama, are relatively small markets.

Mexico

In the early 1990s in Mexico about four of every five stations had a power of 1 kilowatt (KW) or less. Many small, low-powered radio stations cover local areas of Guatemala. Cuba had 115 transmitters, but in many cases the stations were part of a broader network operated either by the government or by a private company. In the case of television, individual station signals, if not part of a network, are extended through repeaters. In Guatemala, for example (where all TV stations are in the country's capital, Guatemala City), TV signals reach much of the rest of the country through repeaters. Generally, the poorest countries rely more heavily on radio than on the more expensive medium of TV. Guatemala, Honduras, and Nicaragua are countries where radio has traditionally been the principal medium.

In Mexico, broadcasting is controlled by powerful business interests. The federal government has the legal right to regulate the privately owned media industry, but in reality its symbiotic relationship with the industry creates a laissez-faire situation.

Mexico's 1960 Federal Law of Radio and Television provides the authority to regulate the media. Its principal purpose is to promote Mexico's cultural and social heritage in the media. Regulatory responsibilities under the law were divided among several ministries, under the tacit coordination of the Secretary of Communications and Public Works. Frequency allocation, licensing, and monitoring of technical standards are under the Secretary of Communications and Transport. The Secretary of Government was assigned the power to monitor programs to ensure adequate promotion of Mexican traditions and family life. The Secretary of Education not only uses the media to improve education, but was given the responsibility of approving announcers at all stations. The Secretary of Health and Public Welfare monitors any advertisements for medicines and food that might be harmful to the public.

In addition, the 1960 law has a number of sections defining acceptable content in programming and advertising, and prohibits the broadcasting of any program that might "corrupt the language and is contrary to the accepted customs" of the country's culture. In effect, if the government wishes to, it can exercise controls tantamount to censorship. Ironically, Mexico, with a Catholic population of about 90 percent, is the only democracy in Central America that bans religious programming. In part, a separation of church and state principle applies; but in great part this reflects the attitude of the people, many of whom believe the Church has in the past supported authoritarian rather than democratic regimes, and fewer than one-third of those identified as church members attend church.

Control of the media has devolved upon the Secretary of Communications and Transport and the Directorate-General of Radio, Television, and Cinema (RTC). The RTC operates Mexico's external foreign radio service, and distributes a weekly radio program and a daily TV program from the government that the 1960 law requires every station in the country to carry. Instead of paying fees or taxes to the government, stations grant up to 12.5 percent of their air time to the government.

While the Federal Law of Radio and Television provides for much stronger direct control than, for example, the U.S. Communications Act of 1934, it is rarely applied. The media industry is led by a private company, Televisa, which owns and operates radio and television stations, for which it also produces programming. Televisa operates under the law, but is so strong and influential that it bypasses it when it wishes. The government dependence on Televisa money includes a joint venture to construct satellite earth stations and joint production of educational programs for public television stations. Televisa is responsible for the development of cable in Mexico, owning the majority of cable systems in the country. When the government established fees and rates for cable systems, it was Televisa that most strongly influenced the nature of the legislation. Imevision is the state-operated network. Some three dozen independent stations and cable systems operate in the northern border area, in major cities, and in tourist centers. To protect Mexican business, these outlets are prohibited from carrying commercials that do not originate in Mexico. The Mexican government also operates an external service and an educational satellite service that serve rural areas of the country.

Honduras and El Salvador

Honduras and El Salvador are classic examples of state-controlled systems. Honduras does not have a detailed, extensive communications policy, but it controls the media in two distinct ways. Through its assignment of broadcasting frequencies and station licenses, it can be certain that any station allowed on the air carries content satisfactory to the government. Second, it maintains dependent relationships with the private owners, some of whom are relatives of high government officials. For example, the principal TV stations and largest radio networks are owned by just a few people who have close ties to the Honduran military rulers. Most stations in Honduras are private commercial operations. As of the early 1990s, Honduras had nine TV stations in operation, plus a number of local stations and cable systems. The cable company, Telesystema, provides subscribers with programming from the United States, such as CNN and HBO. It also had some 300 radio stations on the air. Licensing and control of the stations is under the government office Empresa Hondurena de Telecommunicaciones, or HONDUTEL.

El Salvador, through its former totalitarian president José Napoleon Duarte, created a system that still permits government control: arbitrary withdrawal of a station's license. Although essentially a private system, it does the government's bidding out of fear of reprisals from the military. Both countries import U.S. programming, but since the United States' Central Intelligence Agency (CIA) openly supported the dictatorships in Honduras and El Salvador, much of the population, which has supported democracy movements against their rulers, has reacted adversely to such programming. While most of the stations in these countries are privately owned, there is a diversity of ownership—some religious, some government, some public. In the early 1990s two of El Salvador's six broadcasting networks were government owned, airing principally informational and educational programs. Private-channel entertainment programming is mostly imported from Central and South American countries and from the United States. Other U.S. channels can be seen on the country's cable systems.

Nicaragua

Nicaragua, on the other hand, is an example of what can happen when dictatorships are overthrown. Before the Sandinista revolution of 1979, Nicaraguan telecommunications operated under what was called the "Black Code," which gave the government authority to shut down any station that it believed was disrupting public order. The Black Code was the first media regulation to be abolished when the Sandinistas took over. The television and radio systems, which had been privately owned by members of dictator Somoza's family and by his political and business friends, were seized. The country's privately owned systems were required to register with the Administration of Communication Media and pay annual fees. The former government-controlled private-ownership system was changed to a pluralistic system, reflecting the country's new political and social orientation: a combination of government-owned television and radio stations; privately owned media, including ownership by different political parties; a radio network owned by a public group; and church-owned radio stations. Much of the programming is imported, mainly from the United States, Britain, and Germany, and from Spanish-speaking countries. The government used media principally for educational and social purposes; for example, media played a part in the government's improvement of the literacy rate from less than 50 percent to over 90 percent in the space of little more than a year. However, as time progressed, the United States-backed Contras fought with the government for control of the media, and the government gradually pulled back from its pluralistic approach and used the media more and more to present its viewpoints in opposition to those of the Contras. With its increasing stability in the early 1990s, and the decreased pressure from the Contras, Nicaragua was returning to its pluralistic, democratic approach to media operation under its new center-oriented presidential leadership.

Cuba

Conversely, when Fidel Castro overthrew Batista's totalitarian regime in Cuba in 1959, the government took over control of all stations, which had previously been privately owned and highly commercial. One radio network was said to have broadcast over 1000 commercials every day. Today all television and radio in Cuba is operated by a government agency, Empresa Cubano de Radio y Television. While programming is carefully developed and screened to reflect the country's political, cultural, and educational goals, some TV programming comes from foreign sources, principally Mexico, Nicaragua, and Spain. Viewers and listeners are able to pick up off-air TV signals from Florida and radio signals from a large number of U.S. stations.

Guatemala

Guatemala's telecommunications system is an example of combining some of the democratic approaches of Mexico with the tight state control of Honduras, El Salvador, and Cuba. While most stations are privately owned—over 90 percent of radio and 80 percent of TV—the government maintains strict control over them through its Direccion General de Radiofusion y Television National. One of the networks is programmed by the church with religious content. Although there are many local stations, and repeaters extend the TV signal, the large native Mayan Indian population is generally unserved, in part because programming is in Spanish, not their native language. Some leaders of the continuing Indian revolt against an op-

pressive government suggest that the Indian villages are given no access to media as one means of keeping the people uneducated and uninformed (something Native Americans in the United States have also experienced).

Panama

Political history is clearly a determinant of communications policy. Panama is one of several countries whose governments were changed by the United States at various times, usually to keep in power those parties and rulers who kept U.S. investor companies happy by suppressing labor movements and unionization. For many years, Panama's radio and TV stations were owned and operated by the government. Later, licenses were given to private parties whose loyalty to the government was established. After the Panama Supreme Court ruled such practices unconstitutional, Panama gradually relinquished direct control of the media, except for a national radio station and an educational TV channel, and turned over television operations to qualified private owners without applying a political standard.

Other Central American Countries

Costa Rica relies heavily on TV, with eight major networks, all but one privately owned. Almost all of the programming is imported from the United States and from Spanish-speaking countries. Cable is growing in Costa Rica, with the largest operator, Cable Color Television, serving a number of central communities of the country with some forty channels.

A very small country, Belize, still looks on radio as the principal telecommunications medium. The government owns and operates the radio system, while television is operated by private ownership. Licensing and regulation is under the Belize Broadcasting Authority. The television stations serve seven zones of the country with a modicum of local programming, but mainly with satellite programming from the United States, which is distributed over the air or by cable.

Smaller nations in the Caribbean follow similar patterns, based on the political nature of their respective governments, with privatization, especially in satellite and cable, growing. CARIBVISION plays an important role in program exchange. Puerto Rico, a U.S. commonwealth territory, follows the pattern of the United States, and is subject to the U.S. Communications Act of 1934 and the rules and regulations of the FCC. But its operations reflect its own ethnic needs. Most of the TV and radio stations broadcast in Spanish. Unlike the U.S. mainland, key stations are noncommercial educational or public stations run by educational or religious institutions. In fact, principal news and information coverage comes from the station operated by the Puerto Rico Department of Education and from other public-service channels. The U.S. military operates three radio stations and one television station, broadcasting in English.

SOUTH AMERICA

Third-world conditions in most South American countries result in telecommunications systems similar to those that dominate Central America and other third-world sections of the globe, especially those in Africa and Asia. A major difference in South America, as in Central America, is the influence, investments, entrepreneurial ownership, and technological assistance of the United States. While radio and television are more important in bringing entertainment and information to the public than the print medium is (because of the high rate of illiteracy), radio is the dominant medium because of TV's higher cost. Most systems are privately owned, some with government domination in order to promote the purposes of the

party in power and/or to strengthen the cultural unity and educational uniformity of the country. Where democratic revolutionary movements have been successful, there have been acknowledgements of the people's wish for more freedom of information and media access. Importation of foreign television programs and the gradual growth of co-production has opened up the airwaves to a greater variety of ideas, although almost all imported programs consist of popular entertainment, such as Spanish-language soap operas.

While radio largely began as government-sponsored or -owned stations, television began in most cases under private ownership and, following the U.S. model, became highly and successfully commercial. As in some Central American countries, where the church wields great power and strongly influences government operations, church-operated stations provide an interesting mix of materials that on one hand can be considered ethically oriented and on the other hand are frequently supportive of a totalitarian regime.

Brazil

The largest country in South America, Brazil, also has the most extensive telecommunications systems, affecting in many ways the structures and orientations of its neighbors. Radio began in Brazil in the 1920s and television in the 1950s. In both cases the cost of receivers at first made set ownership prohibitive for the majority of the people, but, also in both cases, the introduction of commercialization made it possible to produce more sets at lower prices. At first government owned and operated, radio (and subsequently television) added privately owned stations. A military coup in 1964 established strict state control over all media. A National Security Law gave the military power to censor all media on the grounds of national security. A Press Law put all media under military control whenever the military wished. A National Department of Telecommunications and a National Council of Telecommunications were established to administer a precoup 1962 National Code of Telecommunications that provided for privately owned stations. Although the military did not directly seize the private stations, its control over them was complete. Both radio and television use grew under the military dictatorship, which lasted twenty-one years. In fact, Brazil became a hemispheric leader in the use of satellites for program distribution, and was one of the first countries in the world to experiment with satellite transmission to get educational programs to the many isolated areas of the nation.

Time was preempted for government programs, specific programs were mandated, and one hour each evening was devoted to a government propaganda program, "The Voice of Brazil." That program, which continued even after the ouster of the military regime in 1985, was so disliked by the people that it was derisively called "the hour of silence." With the restoration of a democratic system it was expected that telecommunications content and access would be much freer. Unfortunately, as of this writing much progress still has to be made. For example, the largest TV operation before the military coup was GLOBO. GLOBO became a highly supportive outlet of the military after the coup, and accordingly grew into a virtual monopoly with government backing. GLOBO still remains the most powerful telecommunications entity in Brazil, now backed by the government and international corporations that find its support of value in their dealings in Brazil. The media do not serve the needs of the many indigenous peoples of Brazil and, by and large, either support or ignore the government's continuing campaign against the country's native populations.

Argentina

South America's second largest country, Argentina, has had a telecommunications history somewhat similar to Brazil's. Beginning in the 1920s, radio grew principally through the development of two privately owned networks. Juan Perón used radio effectively not only to gain power but to solidify it. The government took over all radio stations. In 1951 TV was introduced, also under state ownership, with the assistance, as in many Latin American countries, of U.S. money and technology. After Perón fell in 1955, there was a growth in both government and private stations. Advertising became a key factor for economic support of broadcasting. As TV grew, it began to rely heavily upon foreign programs, mainly from the United States. In fact, the three major Argentine TV stations in the early 1990s were using programs produced by firms that have substantial backing from the three major U.S. TV networks. The growth of media and the political changes since Perón ruled Argentina were reflected in the media policies of a new Perónista president elected in 1989. Two of the three government-run TV channels were sold to private interests, leaving only one public television channel. Although the media operate under an essentially democratic system, they are strongly influenced by three major factors: a virtual monopoly by wealthy media conglomerates, a reliance on advertising from U.S. multinational companies doing business in Argentina, and the continuing possibility, under the law, of government censorship at any time on grounds of national security.

Venezuela

Venezuela is one of the countries in South America that is frequently considered democratic in nature and operation. Its telecommunications history, however, is similar to others in the region. Following the ouster of the military government in 1958, telecommunications reforms were proposed that would result in the establishment of radio and TV networks to serve the public needs of education and information, with restrictions on commercial advertising. The existing private commercial networks defeated such attempts in order to retain a virtual monopoly on Venezuelan telecommunications. Ironically, these private stations are dependent upon some federal funds for their support. The government, however, does exercise a degree of control over content through its Resolution 500, which classifies programs as to their degree of adult content, and which specifies the times such programs may be broadcast. This is somewhat similar to the 1990s indecency rules of the United States' FCC.

Chile

Chile's telecommunications industry still suffers from the reign of Augusto Pinochet, whom the United States installed as dictator when it overthrew the socialist government of Salvador Allende in 1973. Although radio and television stations were privately owned, they were taken over by the military, simply by the government's appointment of new station managers. While the basic laws provided for freedom of expression and for private ownership of the media, various codes permitted government censorship on various grounds. A 1944 broadcasting transmission regulations law continues to remain in effect, but its implementation has depended on who was in power. Radio began in Chile in 1932 and copies the commercial, private-ownership system of the United States, with the exception of a government-operated Radio Nacional network. TV, introduced in 1959, was at first a government operation comparable to the U.S. public television network, oriented toward educational programming.

Although changes were expected with the withdrawal from official office by Pinochet in 1990 and the election of a new president, the continuing control of the military by Pinochet has prevented any substantial telecommunications reforms, and at this writing the control and operations of the media in Chile are little changed. Chile's principal TV network, Television Nacional de Chile, is owned by the government. Four TV stations are privately owned—but by universities, which are subject to government control.

Peru

Peru is another example of the promise of democracy falling to the rule of the military. Radio began in Peru in 1921 and television in 1958. A succession of military rulers kept tight control over the media, and a 1971 General Telecommunications Law officially gave the government ownership of 51 percent of all privately owned television and 25 percent of all privately owned radio stations. With the operation of its own TV network and further control of non-governmental educational TV stations, government power over television was total. While it maintained control, the government did not fully subsidize all stations, and the privately owned stations depended on heavy commercialization for survival. A conservative president elected in 1980 had close ties to the private sector and began a laissez-faire policy, and a new president in 1985 continued that policy. While there are close relations between government and the media, and the government continues to operate its own stations, the wealthy media, as in other countries, supported the government. Nevertheless, in the 1980s, under a new Law of Telecommunications, individual stations began to have more freedom of programming. In 1992 President Alberto Fujimori did what other ostensibly democratically elected presidents have done in other countries throughout the world, as indicated previously: in a remarkable parallel to Russia, he seized control of all the media in the name of national security, following his dissolution of the Peruvian Congress. At this writing, Peruvian law makes it possible for the government to use all of the media as propaganda arms.

Colombia

Colombia has a system that candidly does what some other countries attempt to do in covert ways. The government owns all the television stations, administered through the Telecommunications Division of the Ministry of Communications. However, it leases time to private operators. While at one time there were privately run TV stations, the government ultimately acquired them all. Radio has operated much more freely, with the two largest stations reflecting opposing political views: one liberal, the other conservative. Although a 1959 law guaranteed freedom of the press, it also gave the government the right to censor stations for so-called national security purposes, and even to censor content that the government decided related to family values, police and crime activities, foreign relations, and other matters. While these restrictions have been eased with the election of more democratically inclined officials, national security remains a key rationale for government control.

Ecuador

Despite a U.S.-type system of principally private ownership for broadcast stations, the government in Ecuador has maintained strong control over the media. Following the replacement of military rule by a civilian government in 1979, Ecuador's new constitution guaranteed freedom of speech and communication. But the caveats of national security and national emergency gave the government powers over the media that it has not hesitated to use. Stations have been closed down by the government and, when permitted to operate by the civil government, sometimes have been subsequently closed by the military.

Guyana

Guyana has no clearly stated telecommunications policy, and over the years the government has arbitrarily taken control of what were once private stations. The only news permitted is that provided by the government. In recent years, the new president has begun to allow some freedom for the media. In the early 1990s two private TV stations were operating, but they were restricted to the rebroadcasting of U.S. programs.

Bolivia

Bolivia is typical of Latin American countries in which the media do not serve the nation's indigenous peoples. While most Bolivians speak the Indian languages of Quechua or Aymara, most stations broadcast exclusively in Spanish. This is because most stations are either owned by or influenced through advertising by foreign companies pumping their products to the economically viable Spanish-speaking population. While the government operates its own stations and tries to maintain control over privately licensed stations, its communications rules and regulations are so vague that it legally has been unable to do so. Some stations are even operated by organizations supporting policies inimical to those of the government.

Uruguay

Uruguay is an example of government control from the very beginning of media development. Although when broadcasting began there were no federal regulations governing the media, all licenses were obtained through presidential decree, thus making certain that all stations were loyal to the government. Government censorship grew, and when a military regime took power in 1973, control of the media became total. Any stations not supporting government policy were immediately oppressed and suppressed. The Uruguayan Broadcasting Corporation, under government control, has power over all communications. The return of a civilian government in 1985 augured a change in policy, but the military remained in important government positions and has strong influence on the president, whose right of veto of legislation has prevented any significant change.

Paraguay

Paraguay, similarly, has maintained military control over the media. Since 1954 it has been a military dictatorship. ANTELCO is the country's national communications agency, which regulates on an ad hoc and unchallenged power basis. While there are privately owned stations, they support government policy or risk being temporarily shut down or having their licenses withdrawn entirely, both of which are frequent occurrences.

EUROPE

United Kingdom

While the U.S. system served as the model for many other countries for privately owned commercial telecommunications, the British system served as a model, first, for broadcast media as a public trust and, later, for a combination of public and private stations. As noted earlier, many state systems are based on the structure and organization of the British Broadcasting Corporation (BBC). British television writer Dennis Potter was quoted in the New York Times of October 3, 1993, as stating that the BBC has played a significant part in the lives of millions of people and that "there have been all too few British institutions of any size of which one could say . . . they truly work, that they are the best there is."

The BBC operates three radio channels, appropriately named for their programming as "highbrow," "middlebrow," and "lowbrow." The BBC competes with a number of private radio stations throughout the United Kingdom. Its two main television channels compete with two nationwide commercial channels, as well as with an increasing number of cable and satellite channels. By 1995 the principal competition and programming alternative to the country's four TV networks was a satellite channel, British Sky Broadcasting (B SKY B), privately owned by world media mogul Rupert Murdoch. The BBC also provides radio and TV world service broadcasts, as well as domestic programming.

The BBC operates under an independent board of governors, although because it carries no commercial advertising it is dependent upon the Parliament for its budget. Established in 1922, the BBC received a Royal Charter in 1927 and was put under the Secretary of State for the Home Department in 1981. The BBC is authorized to provide a range of broadcast services, to establish national, regional, and local advisory councils, and to arrange for audience evaluations of its operations. It may not editorialize and is expected to be impartial in covering all issues. While not subject to direct censorship, it nevertheless is sometimes forced to adjust its program content in order to continue to receive financing from Parliament. For some thirty years the BBC operated as a broadcasting monopoly. The monopoly was eventually ended because of concern over one entity controlling the dissemination of news and opinion over the airwaves.

The Independent Television Act of 1954 established the commercially oriented Independent Television Authority (ITA) to provide to the United Kingdom the kinds of programming—essentially LCD popular shows—not offered by the BBC. Two commercial systems under the supervision of the Independent Broadcast Authority (IBA) provide such programming, principally through contracting with independent producers for programs to fill the two channels' time slots. The Broadcasting Act of 1990 established the Independent Television Commission (ITC) and the Radio Authority to assume the previous responsibilities of the IBA.

Ireland

The BBC and ITA serve England, Scotland, Wales, and Northern Ireland. The Irish Republic has a somewhat comparable, but different system. Because the signals from the U.K. are easily obtainable in Ireland, the public has a variety of sources from which to choose. The Wireless Telegraphy Act of 1926 established broadcasting in Ireland, which at first principally relayed radio programs from the BBC. At first broadcasting was a state-operated monopoly, in part because the relatively new republic wanted to use the media to strengthen its cultural and political unity, and in part because some of the early applicants for private stations were thought to be backed by the British. In recent years the existence of pirate stations has prompted Ireland to consider licensing some private stations. Ireland's Broadcasting Authority Act of 1960 established Radio Telefis Eireann (RTE) to develop and "maintain a national television and sound broadcasting service." The first TV channel went on the air in 1961. While the RTE operates with a fair degree of independence, it does remain the administrator of a state telecommunications monopoly. It reports to the Minister of Posts and Telegraphs, allowing the government to exercise a large measure of control over the broadcast media. The principal areas of control relate to censorship of material deemed harmful to national security and, in some instances, not in the general public interest. The government reserves the right to command time on stations for government pronouncements. Although there are license fees for support of broadcasting, the revenue is insufficient to subsidize all the stations and they rely very heavily on commercials. In fact, some critics say that Ireland's broadcasts have the highest percentage of advertising of any country in the world. With the advent of cable and satellite, as well as continued competition from British radio and TV, the de facto audience monopoly of RTE seems to be eroding.

The Continent: Toward Privatization and Diversity

Broadcasting on the European continent reflects not only the diversity of the countries, but the highly competitive nature of their relationships, even with the cooperative programs of the EU. Most countries, including many in eastern Europe since the dissolution of the Soviet bloc, have provided great flexibility in licensing and individual station and distribution system growth. Nations are moving away from what were formerly state monopolies toward increasing privatization. But there has not been a complete turnover to the marketplace. While the private stations are for the most part not controlled by the governments, the governments, as licensing authorities, maintain more regulation than, for example, exists in the United States. The cutthroat competition one finds in totally private systems is widespread, with highly competitive ploys for audiences. As the number of stations grows, individual countries are finding it more and more difficult to compete with over-the-border and satellite signals. In fact, the EU's move toward a unified media system is in serious trouble because of this. Several countries' broadcasting systems have suffered at the hands of outside stations, like Luxembourg's entertainment-oriented CLT and France's TF1. In 1993 CLT alone was operating seven television and thirteen radio stations beaming across Europe. Belgium, for example, was concerned in the early 1990s that its own system might go under because of the heavy competition. Most countries, even those without privately owned stations, have incorporated some form of commercialism as a means of financial support.

The closeness and relatively small size of so many western European countries contribute to the competition. Cable and pay-TV grew tremendously in the early 1990s, especially in Scandinavia, France, and Belgium. The last, for example, had more than twenty-two cable and satellite operations, in addition to broadcasting stations, competing for the attention of a total population of only four million.

While as a whole there is freedom of

the press and of program content, all governments exercise some degree of content control. There is a balance in most instances between the public interest, as represented by an elected government, and the interests of the commercial operators. As Sydney Head stated in his book *World Broadcasting Systems*, European countries have "pinned their faith on the possibility of getting a better wisdom than that of the marketplace."

Western Europe

Germany

Germany is a case in point. Following reunification in 1990, new broadcasting legislation was enacted for the former East and West German telecommunications systems. Radio started in 1923 principally as an entertainment medium, supervised by the Reichpost (PTT). The government maintained strict censorship of news and public affairs. The system's structure, however, was modeled after the BBC. Television was added in 1935. During the reign of the Third Reich, through World War II, the electronic media were used by the state for propaganda purposes with extreme success. After the division of Germany following the end of the war, the occupation powers in the western section, the United States, Great Britain, and France, established regional broadcasting stations in their respective zones of occupation, and the Soviet Union did the same in the east.

The stations in the west became the key networks in the Federal Republic of Germany's own telecommunications system in the 1950s. Both radio and television grew, run by public corporations controlled by citizen groups in the south, north, and west, and in Berlin. There was no private ownership and no profit-making commercialization, although there were limited and grouped commercials on some channels. Business interests pushed hard for privately owned stations with an advertising base, but to no avail. However, in the 1980s they were able to translate their media profit motives into reality with the government's authorization of cable on a private, profit-making commercial basis. The inception of private cable systems set off a media mogul war, with several individual entrepreneurs vying for monopolistic control over private television—in this case, cable—in Germany. Unsavory practices reached such a point in 1990 that renewals were denied for two cable stations owned by the Kirch conglomerate. Privately owned commercial radio stations became a reality throughout the many districts of unified Germany in the 1990s, and this transition to advertiser-driven broadcasting has been a veritable bonanza for outside investors, Americans in particular, as it has in other European countries experiencing a similar conversion.

The three regional state broadcasting systems are the Norddeutsche Rundfunk (North German Broadcasting), Suddeutsche Rundfunk (South German Broadcasting), and Westdeutsche Rundfunk (West German Broadcasting).

The reunification of Germany created an imbalance in telecommunications as well as in other areas of key resources. While the former East German telecommunications system was incorporated with that of the west, its need for updated facilities and trained personnel to match the programming of the west put a financial strain on all of German telecommunications. The opening up of a new area to private cable created a gold-rush mentality for German media barons. A key factor is that Germans in the east watch more television on the average than do the Germans in the western part of the country.

Each regional system has its own board of directors, drawn from citizen groups such as labor unions, educators, religious organizations, and others. While each regional system receives a proportionate share of the license fees collected by the federal government, each operates independently of the federal government. A Coordinating Association of Broadcasting Corporations

(ARD) is charged with coordinating the common interests of the regional systems, such as program production and exchange and mutual legal, political, and economic matters in relation to the central government. In addition, Germany operates both a radio and a television external service to the rest of the world.

German telecommunications is under the general supervision of the Ministry of Posts and Telecommunications. However, in early 1995 German officials met with the head of the U.S.'s FCC to explore changing Germany's licensing and regulatory procedures to something similar to those used in the United States.

France

France's telecommunications vicissitudes have reflected those of the government. Over the years France has jumped from public ownership to private ownership to public ownership and back again, until today it has, temporarily at least, reached a combination of government public stations and private commercial stations. Following the end of World War II, France established state-controlled radio because the pre-war privately owned stations were deemed to have collaborated with the German occupiers and their puppet Vichy government, and therefore were not to be trusted as representing the interests of the people of France. But, as in most governments, radio (and subsequently, television) was used to forward the policies of the party in power. In 1982 a new communications law reorganized France's telecommunications system, including frequency assignments, centralized production and dissemination, advisory councils, advertising rules, and cable authorization. Some degree of local station ownership was allowed, but still under strict government regulation. The Ministry of Posts and Telecommunications is the country's regulatory agency and operation of the several radio and television networks is under the Office de Radio-diffusion-Television Francaise (ORTF). Advertising is permitted on television, but not on radio, with funding coming from the government through set owner license fees. France also operates DBS and overseas services. The state monopoly began to weaken in the 1980s, when a new private TV channel (Canal Plus) was authorized, fiber-optic cable grew rapidly, and satellite reception of foreign channels eroded the government channels' viewing base. By the 1990s France once again made drastic revisions in its broadcast system, selling off a number of channels to private owners, although maintaining regulatory powers over these stations.

The National Commission for Communication and Liberties (CNCL) was established in 1986 to exercise general jurisdiction over the private system of stations. To a degree it operates like the FCC, but adds a program monitoring function and the regulation of political and electoral broadcasting. CNCL is the licensing authority for private stations. A different office, Telediffusion de France (TDF), has jurisdiction over television and radio transmitters, allocating use of its facilities to licensed users. France has two other major offices related to telecommunications: the National Audiovisual Communications Council (CNA) advises the CNCL on public telecommunications, and the Regional Audiovisual Communications Committee advises on the preservation of traditional culture in the different regions of the country.

Belgium

Belgium and Holland are illustrative of small countries that have arrived at almost opposite policies to serve the telecommunications needs of their people. Because of its location and size, Belgium depends heavily on foreign broadcasts, principally spillover signals from adjacent countries. At the same time it has its own broadcast services, one in French through the Institute for Belgium Radio and Television—French Broadcasts (RBT), and one in Flemish through the Institute for Belgium Radio and Television—Dutch Broadcasts (BRT), and a limited radio service in German. The coordinating arm for all broadcasting services is the Institute for Common Services, with the Ministry for Cultural Affairs as the regulatory body. As is also true in Holland, the Belgian audiences seek popular entertainment fare from outside sources, and rely heavily on stations from Luxembourg, whose signals are specifically designed to serve the Belgian and Dutch populations. Belgium's size also has permitted cable saturation, making it the most densely cabled country in Europe, with over 90 percent of it connected.

The Netherlands

The Netherlands has, as noted earlier, the most democratic and at the same time the most complicated system of telecommunications in the world. After establishing the first regular broadcasting station in the world, in 1919, Holland developed a system called "pillarization." Under this system time on available radio and television channels is allocated to citizen groups on the basis of their size, in proportion to the country's population. This allows apparent fair distribution of media access to all citizens, such as labor, religious, educational, and political groups. The channels are not owned by these groups but by the government, and the groups are licensed by the Ministry of Welfare, National Health and Culture. Numbers dictate the eligibility and categories for air time: in the early 1990s category "A" required a minimum membership of 450,000; category "B" 300,000 to 450,000; category "C" 150,000 to 300,000; and category "D" from a minimum of 60,000 to 150,000.

Organizations authorized for air time include the following: the General Radio Broadcasting Association, the Evangelical Organization, the Catholic Radio Broadcasting Foundation, the Netherlands Christian Radio Association, the Workers Radio Amateur Association, and the Liberal Protestant Radio Broadcasting Foundation. One problem encountered by pillarization is "trossification," whereby a number of groups

not necessarily with common interests form a consortium with numbers large enough to gain large amounts of time on the electronic media.

The Broadcasting Act of 1967 established the Nederlandse Omroep Stichting (NOS), which coordinates production and air facilities for the user groups. A Commission for Broadcasting appointed by the Crown oversees the work of NOS. While there is no censorship or prior restraint, the country's Broadcasting Act of 1980 authorizes the Ministry of Culture, Recreation, and Social Work to take post-broadcast action for violation of rules concerning indecency, national security, and advertising.

As in other countries, satellite and cable have challenged the broadcasting base. Next to Belgium, Holland has the highest cable penetration in Europe, about 90 percent, and as noted earlier, has a large audience for programming from Luxembourg—which is actually produced in Holland—for Dutch audiences. In 1993 the state-owned telephone company, which owned a significant part of the country's cable systems, was faced with greater privatization, including international investment in, and operation of, Dutch cable. DBS is also growing.

Advertising provides about 25 percent of the budget and is restricted to certain blocks of time. The rest of the budget comes from user license fees. Advertising is not handled by the user groups, but by the government's Broadcasting Advertising Foundation, Stichting Ether Reclame (STER).

Scandinavia

Broadcasting in Scandinavia is undergoing painful change. Established as state systems to serve the cultural and educational needs of the public, with limited government control and virtually no censorship of any content, telecommunications systems in Sweden, Norway, Finland, Denmark, and Iceland became known for high-quality programming. While some countries allowed some private ownership over the years, others did not, and the systems by and large became public entities under government supervision with government support. In some cases the private stations operate without a commercial base, but with government funding from user license fees. Principally, broadcast systems in the Scandinavian countries operate in much the same way that the British system does. In Sweden, the Swedish Broadcasting System (SBC) supervised the physical and transmission facilities, handled the funding, and allotted air time, while a public nonprofit corporation under an organization called AB took care of programming and production. In 1993 the broadcasting system was reorganized into three independent organizations: Sveriges Riksradio Ab, Sveriges Television Ab, and Sveriges Utbildingsradio Ab, the last responsible for external programming. While all stations are under the government, legislation was being considered that would authorize private radio stations. The technical transmission facilities and budget allocations from license fees are under an independent agency, the Swedish Telecommunications Administration.

The same is true in Finland, where the Finnish Broadcasting Corporation (Oy Yleiradio Ab, or YLE) functions with government funding, like the BBC. While YLE operates the three national radio and two national TV networks on the air in 1990, it shares control of a third, private, network, principally through its operation of the facilities. However, in 1990 YLE began losing more of its control over TV to some fourteen private stations that were being distributed via cable. Satellite broadcasts are also growing, providing serious programming competition to YLE. In the early 1990s YLE was distributing its TV and radio signals through some forty TV stations and thirty TV boosters, and some sixty radio stations. While privatization is developing in Finland, the government and YLE have attempted to maintain control, if not dominance, by becoming partners,

if only through rental of facilities and air time, to private broadcasters.

Danmarks Radio is Denmark's equivalent of the United States' FCC, regulating, licensing, and establishing broadcast standards, but it has the added responsibility of allocating funds to stations from the collected user license fees. The government controls all broadcasting, with the TV and radio networks operated by Danmarks Radio. Although advertising is permitted on some stations, it is restricted by time and placement limits. Cable systems are operated by private companies, with satellite programming included in their program fare.

The Scandinavian countries have jealously guarded their cultures and were much slower than other European countries to develop satellite and cable facilities. However, in the 1990s the spread of DBS lured many viewers away from the state systems to foreign popular entertainment, and several Scandinavian systems were reorganizing and retraining their personnel to develop competitive programming that would still retain their former quality and cultural purposes. At the same time, there is a trend toward decentralization and the authorization of more private stations.

In Finland the broadcasting system is governed by a board appointed by the Parliament, proportionately representing the various parties' parliamentary numbers. In Norway a dual system exists, a board appointed by the Minister of Culture and Science Affairs and a council appointed by the Parliament. In Denmark the members of the governing broadcasting board are appointed by three bodies: the Ministry of Public Affairs, the Ministry of Culture, and the Parliament. In Sweden the council governing the Swedish Broadcasting Company's three successor groups is appointed by the government, and in Iceland broadcasting is under the aegis of the Ministry of Education.

Scandinavian countries held out longer than the rest of Europe against commercialization. In recent years economic exigencies have led to the inclusion of advertising on some channels, but it is carefully controlled as to type and length and is restricted to designated time blocks. The principal source of funding in all Scandinavian countries is user license fees.

Southern Europe

Southern European countries show some similarities in their system types and controls. Italy, Spain, Portugal, and Greece have all used the media to forward the policies of their respective governments and, in order to do so effectively, have maintained some degree of government monopoly through ownership or strict control of the media. As satellites and cable brought foreign broadcasts to the public, these countries were forced to follow the same path as their neighbors to the north and move toward more flexible systems, with private ownership and local stations.

Italy

Italy, for example, established its government communications monopoly in 1924 with its first radio license, issued to the Italian Radio Union (URI). Subsequently, the fascist government took even firmer control and established a state-run system, the Italian Radio Audition Corporation (EIAR), which was later replaced by the Italian Radio Audition (RAI), which continues today. The state monopoly was reinforced by the Italian courts in 1960 on the grounds that it was the government that knew best how to serve the majority of the people and that the government would be more objective and impartial than a private owner. However, the public objected to the ruling conservative party, the Christian Democrats, being the arbiter of what was objective, and in 1975 new communications regulations made the state system into a BBC-type organization, operated by a public corporation, and allowing a number of local publicly controlled stations. With the end of the state monopoly, private stations sprang up throughout the country. While freedom from government control was the rationale given for the change, the increasingly powerful business interests in Italy were seeing too much potential advertising profit slipping away, and their pressures were also responsible for the change. Various groups, ranging from the church to labor unions to political parties to purely private profit-making companies, established their own stations and in 1984 were given permission to establish national networks. Private entrepreneurs began to monopolize the telecommunications system, replacing the government in this role. The government system, RAI, was beginning to suffer, and in 1990 the government passed an anti-monopoly law that restricted any one owner to 25 percent of the national networks. However, two major companies were awarded licenses that virtually shut out many of the smaller independent operations. One conglomerate is owned by media mogul Silvio Berlusconi, who in 1994 parlayed his media power and exposure into election as head of the Italian government.

Italian broadcasting is controlled by several duopolies. Private stations garner more than half of the nation's audiences. RAI—Radiotelevisione Italiana —is governed by a parliamentary commission representing all political parties. This commission, in turn, appoints a Board of Administration for everyday supervision. RAI's three radio and television channels are technically given national network monopolies, but in fact seven private national networks challenge them directly.

Two federal agencies are in charge of telecommunications in Italy. The National Private Broadcasting Supervisory Commission functions as a regulatory body, somewhat like the United States' FCC. The National Broadcasting Committee, under the Ministry of Posts and Telecommunications (PTT), is the licensing body. The growth of cable prompted tighter regulation in 1993, with cable systems for the first time obligated to apply for licenses from the PTT.

Spain

Spain underwent a similar change from a dictatorship to a democratic form of government. However, its government in recent years has been more stable than Italy's, accounting for greater stability as well in its telecommunications system.

Broadcasting in Spain was originally established as a state monopoly, although in principle there was authorization for private stations. However, the government always maintained strict control, and in Spain's Radio Broadcasting Act of 1934 government monopoly was reaffirmed. The Franco regime, from 1936 through the dictator's death in 1975, totally controlled the media, and established the Spanish Broadcasting Authority, which came under the jurisdiction of the Ministry of Education. Franco did authorize some new stations after World War II, several of which were not government operated but were licensed to the government-approved national labor union, private business organizations, and political branches of Franco's party. A large number of stations were authorized for the church. In 1960 Franco abolished all church stations, concerned that the church might not continue its strong support of his totalitarian regime. In 1977, after Franco's death, a new decree proclaimed the freedom of broadcasting, but, in fact, the government merely reorganized the system and kept it under government control. The new Spanish constitution of 1978 reaffirmed such control, although in subsequent years limited private ownership was instituted and regional channels came into being. In 1980 a public corporation, RTVE, was given authority to oversee all broadcasting in Spain. Private broadcasting was ruled legal by the courts in 1982, and by 1990 three new TV networks were in operation. The Broadcasting Law of 1989 authorized local TV stations and

cable systems. In the early 1990s 1400 new radio stations went on the air. The Junta Nacional de Telecommunicaiones, under the Ministry of Transportation, Tourism, and Communications, is the principal regulatory agency, and assigns frequencies and licenses. Program content is overseen by a national Censorship Board. While private broadcasting is relatively unrestricted, the government requires fees from the private stations for use of the government-controlled transmitters and booster stations. The private TV and radio stations, specializing in light entertainment, have become the most popular stations in Spain. In addition to national networks, Spain has regional networks and stations.

Although Spain continues to exercise strong control over its telecommunications system, it is moving toward a more flexible expansion. In the early 1990s, it began to issue licenses for satellite entertainment channels, with a limit of one to any given operator. And in late 1994 the Spanish cabinet adopted a law permitting international investment in and operation of a planned nationwide cable system.

Portugal

Portugal's recent political history is similar to Spain's, and so is its telecommunications history. In the 1930s the government established a state-run broadcasting system, which was slightly modified in 1955 when Radio-Televisoa Portuguesa (RTP) was set up by the government to incorporate private investment into the government stations, adding limited commercial advertising. After the long-time dictator was deposed in 1974, changes were expected in broadcasting, just as they had been in Spain. As in Spain, little changed. A government monopoly was again instituted. With the growing public dissatisfaction with government station programming in the 1980s (a reaction also found in many other European countries), the government began, in 1988, to license some private stations. Also, as in other countries, these stations concentrated on imported popular entertainment programs. Portugal's entry into the EU in the 1980s spurred private station growth.

The Ministry of Industry and Technology establishes the technical standards and requirements for telecommunications in Portugal, while the RTP itself is governed by a separate five-person board.

Greece

Also fitting into the general description earlier applied to southern European telecommunications operations is Greece. The frequent changes from military regimes to democratic governments seem not to have changed telecommunications policies to the extent expected until the early 1990s. The state system of broadcasting established in 1936 included a provision for private stations, but the provision was not enforced. Greece's 1975 constitution clearly put "radio and television . . . under the immediate control of the state. . . ." In 1980 the government established Greek Radio and Television (ERT) to oversee all broadcasting operations. In 1988, in part because of the existence of many pirate radio stations, the government authorized private radio ownership and a year later private television authority, but limited any one owner to 25 percent of the stations. A National Broadcasting Council was created to coordinate the new system. While Greece was moving toward more privatization in the early 1990s, its progress was slow and deliberate; it apparently wanted to make certain that the government retained sufficient control to protect government politics.

Another important comparison among the southern European countries that recently had harsh dictatorships and now have democratic governments is that even with the retention of a great deal of government control over the media for political purposes, each has established governing boards that are more or less representative of the body politic. Further, each has moved toward some degree of privatization of stations. All have strong principles of government involvement, but as yet none have clearly defined sets of rules, resulting in weak or confused regulation. Finally, all have moved in recent years toward the new technology of DBS, as opposed to the cable domination of some European countries further to the north. Satellite importation of foreign programs is expected to speed up each of these countries' integration into the mainstream of international economic and cultural exchange.

Other Non-Eastern Countries

A few other countries in western and central Europe merit mention here as typical of a number of other systems, even though they are dissimilar historically and politically.

Switzerland

Switzerland plays a neutral telecommunications role, serving all the countries in its region, in the same way it has played a neutral political role. The Societe Swiss de Radiodiffusion et Television (SSR) is the Swiss broadcasting corporation that coordinates the three national networks, which serve, respectively, the French-speaking, Italian-speaking, and German-speaking regions. Because Switzerland does not have a specified communications law—other than a general concept of media freedom for the principal purposes of information and education, as stated in an article of the 1984 Swiss constitution—the SSR operates flexibly, along the lines of a public corporation such as the BBC. A national advisory group represents the country's different regions. Attempts in Parliament to create stronger legal controls over broadcasting have thus far been unsuccessful. Switzerland also has a number of privately owned stations that operate on a commercial basis, but with advertising limits imposed by the govrnment. High cable penetration adds to the international commercial aspect of Swiss media.

Austria

Broadcasting in Austria began as private enterprise, but when the country was occupied by Germany in 1938, all radio became a function of government. At the end of World War II the Allied occupying powers temporarily took over the broadcast system, turning it over to the new independent government in 1955. The government maintained tight control over content, but the public reacted adversely to government use of the airwaves for political purposes—to many it was too reminiscent of the years of propaganda use by the Nazis. In 1964 the Austrian Broadcasting Corporation (ORF) was formed to be a buffer between the broadcast system and political influence. The new system was modeled on the BBC. Although much of ORF's funding comes from commercial advertising and private donations, it is dependent for some of its support on the government, which collects and distributes user license fees, thus continuing some government influence.

As a public corporation, ORF is controlled by local governments. This guarantees local participation in its programming and operations. The Ministry of Transport's Post and Telegraph Administration was designated to oversee transmission facilities. In 1974 ORF was legally authorized to operate as a broadcasting monopoly. While broadcasting is under the effective control of the federal government, the country's provinces (Lander) are minority partners and cooperate by providing programming from studios in their regions. In the early 1990s the government was considering the authorization of privately owned stations for the nation's telecommunications system.

Former Yugoslavia

The telecommunications system in the former Yugoslav republic, at this writing

shattered into internecine warfare, provides an interesting commentary on the connections between the communications and the political structures of a country. Under the Yugoslavia Communications Act of 1965, radio and television were decentralized so that each of the six republics and two autonomous regions had its own broadcasting system. Although a national coordinating body known as Jugoslovenka Radiotelevizija (JRT) dealt with general policy and technical operations and collected and distributed the user license fees, the systems operated quite independently, each serving the nationalistic and patriotic purposes of its regional government. Each republic and region determined program content on the basis of its people's interests and needs. Even so, the country as a whole had been active in international matters, participating in the EBU and the IRTO, and operating an external service. Given the Bosnia–Croatia-Serbia conflict of the 1990s, one cannot help but wonder whether the fragmenting of the telecommunications media into regional and ethnic blocks facilitated the subsequent warfare among the different ethnic and religious groups. Would a greater centralization of telecommunications, emphasizing commonalities and unities, rather than xenophobic differences, have changed the course of political events?

Eastern Europe

With the rapid changes in eastern European governments in the early 1990s, and the lack of comprehensive telecommunications laws and/or policies in most of the countries, it is difficult to specify individual countries' systems of regulation and ownership. All of the eastern European countries had for decades operated tightly controlled state systems, either as arms of government agencies or under the direction of the respective countries' Communist Parties. Local stations functioned similarly, being responsible to local committees. Soviet bloc broadcasting systems were further coordinated by a council of ministers representing the various countries. Transmitting apparatus and technical facilities were usually under a Ministry of Communications or its counterpart. As noted earlier, some countries extended the signals to more people by means of repeaters and wire.

Following the velvet revolutions of 1989 and 1990, a number of communications systems were removed from state monopolies, and private ownership was permitted, relying heavily and for the first time on commercial advertising. In some countries user fees were instituted. Previously, the broadcast systems had been subsidized by their respective governments as government-operated entities.

Privatization has created political problems in some countries. In 1993, when the Czech republic granted its first private TV license to an American investment group for a twelve-year term, it precipitated a crisis in the Czech government. Discussions of private licensing in Poland and Hungary have created similar furors. Although newly democratically elected, it is difficult for the officials of the eastern European countries to conceive of a media system that is free to criticize their politics and actions. On the other hand, a 1993 study by the European Institute for Media Studies, stated that there is concern in eastern Europe over "attempts of governments to establish one party controlled television."

Countries in eastern Europe, as elsewhere, are greatly concerned about permitting any foreign ownership of any part of their telecommunications systems. Yet they face the problem of having insufficient funds to develop the systems by themselves and they need foreign investment. Another concern is that private stations will have to rely on advertising for their support, a prospect that disturbs not only governments, but also the viewing and listening public. At this writing, however, it appears that private stations that will operate under a dual system of ownership will be introduced and grow in a number of eastern European countries.

Russia

As noted earlier, Russia epitomizes the dichotomy facing many countries. Even before the change from Communist Party rule in Russia, the concept of private broadcasting was established through Moscow Echo Radio, or "Radio M" as it was called by the public. This independent station presented news and information that raised the kinds of issues previously unaired in Soviet media. The new Russia quickly established the principle of private stations—that is, stations not under a federal government monopoly. But, as we have pointed out, the old ways die hard and when it suited his political purposes, the elected president of Russia, Boris Yeltsin, seized control of the government stations and exercised restraints over the private stations. Oleg Onoprienko, a producer for Radio Rossija, which broadcasts to some 80 million people throughout much of Russia's vast area, wrote to one of this book's authors in the fall of 1993, saying that he "realizes the greatest role free and democratic means of information, including radio" can play, while at the same time he laments what he perceives as a continuing monopolistic control over communications by the state. He attributes this to "present political statesmen's playing on the difficulties of their people."

Russia's radio and television are under the direction of Ostankino—formerly Gosteleradio until renamed by President Boris Yeltsin—and Russian Television and Radio (RTV). They are run by a combination of professionals and those appointed or approved by the party in power, thereby giving the government effective control over operations and content, even when the system is not under direct governmental control. In 1995, as shall be discussed, serious changes in this system were proposed.

All the technical facilities for telecommunications in Russia are under the supervision of the Ministry of Communications. This ministry is also responsible for issuing broadcast licenses. A Federal Information Service oversees broadcasting operations. In 1994, Russia established a Committee for Informatization Policy, under the direct control of the president, which strengthened state jurisdiction over the media.

President Yeltsin's 1992 decree "On Communications in the Russian Federation" allowed any government or non-government entity, domestic or foreign, to set up and operate telecommunications networks in Russia. However, an operator must first obtain a license from the Ministry of Communications. In addition, it must comply with strong anti-monopoly regulations. In 1991, the Ministry of Communications' broadcast License Commission accepted applications for a limited number of private, independent stations. In 1992, the U.S. Turner Broadcasting Company received a joint venture license with the Moscow Independent Broadcasting Company. By 1993, sixty private broadcasting licenses had been issued—not a large number for a country the size of Russia.

However, in 1993 the Russian Federation's Commission on Broadcast Licensing was dissolved, making it possible for many stations to go on the air at will and threatening to create an interference "Tower of Babel" similar to that which forced the United States to enact a Radio Act in 1927. One result is the proliferation in the mid-1990s of stations on a new "western-style" FM band. These are commercial stations, many fueled by foreign investments and influence, and with predominantly western-type pop music.

As the attempts at instant capitalization (as opposed to a gradual transition from a socialist economy) devastated the economic and social fabric of Russia in the early and mid-1990s, privatization in the broadcasting industry suffered the same schizophrenia. While on the one hand committed to supporting privatization of television and radio, on the other hand the government didn't want to lose control of the propagandistic power of the electronic media, and continued to switch back and forth in policy and action regarding control.

In the broader area of telecommunications beyond broadcasting, such as computer, telephone, and new technology networks, Russia moved ahead at a rapid pace in the 1990s, with the government maintaining at least some control over each enterprise. This is in part due to the lack of a substantive number of definitive rules and regulations regarding telecommunications, leaving the government the flexibility of determining its actions on a case-by-case basis.

One of the results of Gorbachev's glasnost and perestroika, however, is that the people have been more open and quick to criticize government censorship of the media than they formerly were, placing pressure on the government to relinquish some of its newly seized dictatorial powers. Under Yeltsin Russia's 1990 Law of the Press and the Mass Media eased state control over the media, prohibited censorship, and provided for public access to the media. However, the law never really reached the stage of full implementation and as this is written is for all intents and purposes suspended.

All eastern European electronic media suffer from a lack of adequate funding, which results in antiquated equipment, limiting both the quality of the programming and the transmission to some parts of individual countries. In fact, in 1994 Russia was seriously considering granting licenses to foreign companies, with the expectation that this would provide needed financing. It was also expected that any such licensing would be accompanied by restrictions that would ban content critical of the government.

In 1995 one attempt at privatization resulted in tragedy. President Yeltsin appointed a new head of Ostankino, the state-owned television system, to reorganize it so that part of it would be sold to private investors. The new head, a popular national TV personality, announced that he would impose a moratorium on stations' advertising, $30 million a year of which was being skimmed off and not reacing Ostankino itself. Instant capitalism in Russia, however, had resulted in chaos and crime as free enterprise ran rampant, and the new director was immediately murdered. Months after, the murder still remained unsolved, and serious reform or movement toward a democratically regulated telecommunications sytem serving the public interest seemed even more in doubt.

Other Eastern European Countries

Poland's radio and TV systems are extensive, and are operated by the government. However, government restrictions on content, plus unofficial but effective control over content by the Catholic church, has prompted efforts to provide alternative broadcasting to Television Poland (TVP), with some of the best journalists now working for TV Solidarity. Hungary exemplifies the rationale of some other east European countries not yet ready to accept some of the freedoms of democracy. Magyar Television is the public corporation serving the people with public-service programming. Using the excuse that the station is being run by former Communists and other left-wingers, the Hungarian government proposed in 1993 to take it over. Many in Hungary saw this as a blatant attempt by the government to seize control of the media, which is prohibited under the new Hungarian constitution. On the other hand, private ownership seemed about to expand in the early 1990s. In 1989, just before the government's ban on licensing commercial stations, the country's largest bank and postal service began operating the first private TV station. In 1993 a substantial part of the system was bought by an up-and-coming Hungarian media mogul, giving rise to the expectation that more private stations would be on the air over the next few years.

Lithuania is typical of the Baltic countries' movement away from Soviet-controlled media to media controlled by the individual nations. The Lithuanian Ministry of Telecommunications and Information supervises the state television and radio companies, which operate the principal systems in the country. Following the declaration of independence from the U.S.S.R. in 1992, private stations emerged. While there is considerably more flexibility than in the past, the state still maintains a watchful eye over content that it fears might harm the new political purposes of the government. Romania is typical of the poorer countries that continue to depend on government control of the media to unify public opinion in favor of the new government. The limited broadcasting system in Romania is run by the state.

Russia still maintains a communications influence, if no longer a direct political influence, on many eastern European countries. Because some countries, such as the Baltics, cannot afford to produce or purchase all the programming they need, they depend on programming from Russia. This results in a nationalistic dichotomy: after years of Soviet domination, many of the people who want the kinds of programs Russia offers resent receiving them in the Russian language.

An interesting side effect of having public and private stations free of government control and censorship is the change in the listening habits of many people. Heretofore, many residents of eastern Europe listened avidly to external services from the United Kingdom, United States, and other countries for news and information not usually found on their own stations. With more freedom of media content, they are now turning to their own stations and away from foreign services.

Eastern European countries are in the throes of determining the kinds of media systems they want and what kinds of regulations should govern them. A 1992 project of the Annenberg Washington Program in Communication Policy

Studies of Northwestern University and the Trans-Atlantic Dialogue on European Broadcasting centered on the systems in the Czech Republic and Slovakia. The questions they raised in the study "Building Democracy: New Broadcasting Laws in Eastern and Central Europe" are the questions being asked in all eastern European countries. Among these questions are the following: should the system be a state monopoly or private or both? what are the criteria for access to the air? should governance of stations or systems include representatives of the public at large? to whom do the frequencies belong—the federal government, regions or states or local governments, or private entities? how will stations be funded and should advertising be allowed?

AFRICA

Throughout Africa telecommunications are owned and operated by the respective governments. In the relatively few instances where private ownership of broadcasting stations is permitted, either the lack of an economic base has prevented such private stations from going on the air or, where they have, the government has retained strict control over them. In some cases, as in Cameroon, even though there is authorization for private ownership, no infrastructure was developed to provide for the licensing of such stations. Even in countries such as Algeria, Ghana, Malawi, Mauritania, and Nigeria, where the media systems are public corporations and usually modeled after the BBC, government policy and pressure ensures that the stations represent the proper political perspectives. Where systems are decentralized, the regional or local government exercises the same kinds of controls that a national government would.

There are several major reasons for this stricture. First, most of the African countries were under the control of colonial powers until relatively recently, with a history of all media having been under colonial government rule. The vestiges of such authority have continued. Second, because most African countries are desperately poor, it is usually only the government that can afford to finance telecommunications systems or stations. Third, the conditions in most African countries provide a rationale for centralization of the media. The high rates of illiteracy, extreme poverty, starvation, death rates about twice that of western nations, and continuing high birth rates make the media a necessary means of education for economic and cultural development.

Not only are the media used by many governments for national unification, for example, as a means to try to bring diverse tribes together, but as a logical tool for intercountry communication to promote Pan-Africanism.

Let's look at a few countries in particular. Ethiopia owns and operates its stations through the Ethiopia Broadcast Service (EBS). However, although EBS operates as an independent public authority, the Ministry of Information maintains strong influence on programming. In addition to national media services, Ethiopia provides an external service in a number of languages. In Kenya, broadcasting is under the control of the Ministry of Information and Broadcasting, which functions as a propaganda arm of the government. In Tanzania, the Ministry of Information controls the media and makes sure that the people receive educational materials that reflect the policies of the government. Although Nigeria has a fairly decentralized system with stations in various states of the country, both the Federal Radio Corporation and the Nigerian Television Authority are ultimately responsible to the Ministry of Information. Libya's broadcasting structure in theory provides for public participation in broadcasting, but in fact the central government maintains total control, a situation common to other countries as well. The Peoples Revolutionary Broadcasting

Company is run by a committee of the General Peoples Congress. But policy and funding are under the direction of the Ministry of Information and Culture.

An incident in Ghana in late 1994 was illustrative of the problems and practices in most developing countries, not only in the region but in all parts of the world, in relation to control of broadcasting. After almost a year in which he was unable to obtain a government response to his petition to start a radio station, one broadcaster simply went on the air with an FM station under the government's constitutional guarantee of "freedom of the press." The government sent armed soldiers to seize the transmitter and shut down the station. The editor of a Ghanian magazine stated that the government leaders "have seen what free electronic media can do in other countries and are scared to death." The government replied that it is "fully committed to freedom of expression," but for the immediate future "radio and television should serve as the forum for a discussion of our development needs . . . that our society is not as stable as some others, and we could have disorder as a result of the misuse of this resource."

Earlier discussions of media in Africa showed how the political systems of the countries, for example in the Gulf States, were reflected in the control of the media. Egypt, with the largest and most influential broadcasting system in Africa, maintains a content-controlled operation despite a government structure as democratic in principle as any African nation. Its operations have been influenced greatly by western communications systems, including the development of training courses for some of its personnel. Egypt's approach has influenced those of a number of other African countries, such as Ghana, Nigeria, and Kenya. The Mediterranean countries of Algeria, Libya, Morocco, and Tunisia have generally followed Euro-American structural patterns.

Of other mideast countries, Lebanon is the only one with a principally private commercial system. But it still remains under government control through the licensing procedure and threats of suspension of operations. The other countries' systems, as earlier noted, are government operated. It is anticipated that the formalization of peace between Israel and its neighbors will open the way for greater entrepreneurship, with the possibility of some private operations in Israel and Jordan. However, that would probably occur only after state stations—for example, a Jordan-aided and perhaps -operated Palestinian station in the West Bank—have achieved their political purposes. There seems to be no immediate prospect, however, that even with the introduction of some privatization, the governmental control over stations, especially their content, will be in any way eased.

The situation in many of the mideast Arab states, in both Africa and Asia, is illustrative of the problems facing some of the third-world countries. Many of them have invested their monies in the technology, with studios and equipment coming first and a concern for adequately trained personnel and quality programming coming second. In many cases, especially in those countries with state religions, programming has a narrow purpose, and the principal aim is to reach as many people as possible with the government's ideas and ideals.

In fact, some systems are simply carry-overs from Euro-colonial days, in many cases using the equipment left behind by the colonial powers. Kenya is an example of such a situation. When Kenya gained its independence, the British left it a government-funded broadcast system, privately owned by a consortium of investors from the United States, Canada, the United Kingdom, and some east African countries. The new government did not have the funds to continue the private operation in the same way, and took over the Kenya Broadcasting Corporation, renaming it the Voice of Kenya. It was put under the Ministry of Information and Broadcasting. As a government nonprofit entity, it could be operated on a smaller budget, and served the government purpose of providing education and information that solidified national unity and supported government policies. This sequence of events occurred in some other formerly colonized nations in Africa, too, with government-run telecommunications systems under strong one-party rule.

Part of the difficulty that some countries have in keeping content under government control is the relatively easy reception of European signals in countries on the Mediterranean coast and the cross-border transmission pickups in most other countries. Further, the public demand for more entertaining programs has led to a growing number of satellite-receive stations throughout Africa, with foreign programming largely bypassing government censorship attempts.

ASIA

Because most Asian nations, like African ones, are third-world countries, strong government control of the media is a common part of national policy as a means of unifying the public, for purposes of economic progress. Asian telecommunications systems range from those with minimal government control, to systems of public corporations and private stations as in Japan, to vast national government owned-and-operated systems as in China. As in other parts of the world, there is not necessarily a correlation between the form of government and the type of media system and its control. Some democracies, particularly those of emerging third-world countries, maintain strict government controls, more so than some countries with monarchial or colonial rulers, such as Hong Kong. Although the Philippines operates within a democratic structure and its media are privately owned, it maintains strong government regulation of broadcasting. Countries such as China and North Korea, whose socialist philosophies differ dramatically from

the neo-fascist principles governing countries such as Iran, Iraq, and Syria, nevertheless maintain the same kind of strict government controls over the media.

Earlier we discussed the mideast countries located in Asia, noting the gradations of control and types of systems, from Israel to Jordan to Syria to Iran to Iraq to Lebanon. Only in Lebanon are there privately owned stations, but these remain under strict government supervision. While a democracy such as Israel may provide more freedom of content for its radio and television stations, it nonetheless maintains ownership and operation of them, as do presidential dictatorships such as Syria and Iraq. Despite the fact that Iran, Iraq, and Turkey have systems that are ostensibly insulated from government interference by virtue of being public corporations, the day-to-day operations are dictated by the government. Systems owned directly by the state, as in India and Jordan, ironically appear to be more independent in operation than the aforementioned public corporations.

Japan

Japan's dual system of broadcasting in the early 1990s consisted of more than 100 TV stations nationwide, licensed as commercial stations and public stations; the networks operated by the government's BBC-like Nippon Hoso Kyokai (NHK), the Japan Broadcasting Company; and stations operated by provincial departments, principally oriented to education. Hundreds of transmitters and thousands of relay stations carry signals to all parts of the country. In the early 1990s a domestic satellite was transmitting signals to northern Japan (Figure 3-1). A fine line is drawn between commercial and public stations. Both types are permitted to carry commercials. Public stations got that name because they ostensibly carried programs oriented more towards the cultural, educational, and informational needs of the public than the entertain-

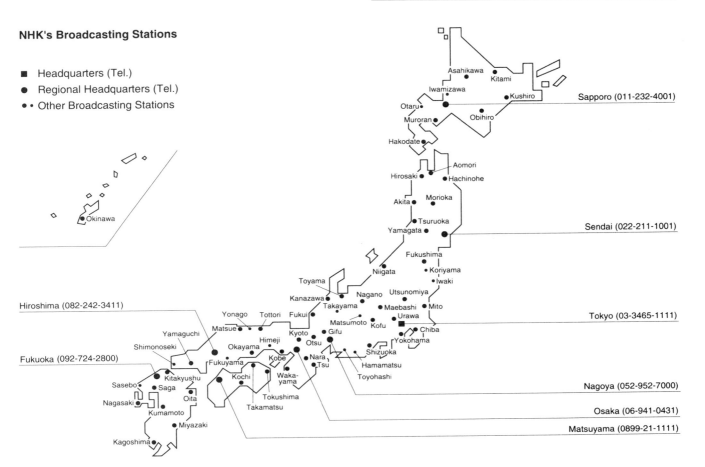

NHK's Broadcasting Stations

■ Headquarters (Tel.)
● Regional Headquarters (Tel.)
● • Other Broadcasting Stations

Asahikawa
Kitami
Iwamizawa
Otaru Kushiro Sapporo (011-232-4001)
Muroran Obihiro
Hakodate
Aomori
Hirosaki Hachinohe
Akita Morioka
Tsuruoka
Yamagata Sendai (022-211-1001)
Fukushima
Niigata Koriyama
Toyama Iwaki
Kanazawa Nagano Utsunomiya
Takayama Maebashi Mito
Yonago Tottori Fukui Matsumoto Kofu Urawa Tokyo (03-3465-1111)
Yamaguchi Matsue Kyoto Gifu Chiba
Himeji Otsu Yokohama
Shimonoseki Okayama Shizuoka
Fukuyama Kobe Nara Hamamatsu
Sasebo Kitakyushu Kochi Waka-yama Tsu Toyohashi
Nagasaki Saga Oita Nagoya (052-952-7000)
Kumamoto Takamatsu Osaka (06-941-0431)
Miyazaki Tokushima Matsuyama (0899-21-1111)
Kagoshima
Okinawa
Hiroshima (082-242-3411)
Fukuoka (092-724-2800)

**FIGURE 3-1
NHK broadcast
sites in Japan.
Courtesy NHK.**

ment programs of the commercial channels. In practice there is sometimes little discernible difference. All commercial stations belong to the Association of Commercial Broadcasters, which operates as a self-regulating organization in order to preclude strong government regulation. It is noteworthy that many commercial stations look upon NHK as the ideal operation and attempt to emulate it, thereby promoting a generally high level of responsibility in Japanese broadcasting. The pre-World War II NHK served the government as a propaganda agency, and following the war the U.S. occupying administration created a new NHK and, in a declaration guaranteeing freedom of speech and press, paved the way for the introduction of privately owned stations. A Broadcast Act was enacted in 1950. Government regula-

tion is somewhat equivalent to that in the United States, with no overt censorship. But because all stations are licensed by the Ministry of Posts and Communications, and NHK relies on the government for its annual budget, the government is able to exercise pressure when it feels it is necessary.

China

Radio began in China in 1923 with the establishment of a private experimental station in Shanghai by an American reporter, E.C. Osborn. The first official station was in Harbin in 1926. In 1928, the first Central Broadcasting Station went on the air in Nanjing, and a year later the government authorized private stations. Under Japanese occupation, Chinese radio became state controlled.

With China's independence the ruling Communist Party established a totally government-run system. Television was introduced in 1958 in the capital, Peking, now Beijing. Television's operations are under Central China Television (CCTV) and radio under China People's Broadcasting Station (CPBS), both under the Ministry of Radio and Television. While the central networks and individual stations do have some autonomy in the development and production of programs, they must be sure that the program content reflects the purposes of the government, not only as delineated by the Ministry of Radio and Television, but as more specifically defined by the Propaganda Bureau of the Chinese Communist Party Central Committee.

Since 1984 China has made extensive use of satellites to provide nationwide coverage of its television signals, used principally to raise the level of formal education and for cultural and political unification.

Hong Kong

In 1997 Hong Kong will become part of China once again, as will its telecommunications system. At the present time it is a mix of public and private ownership. Radio Television Hong Kong operates several government radio and television networks. Hong Kong Commercial Broadcasting Company runs several radio networks, and two private television operators run TV networks. Although ostensibly a democracy, the government's licensing of all stations includes a proviso that only government-issued news may be carried by these stations. Hong Kong audiences, however, have been able to bypass the restrictions on news through the government-sanctioned operation of an extensive cable and DBS system. All of telecommunications in Hong Kong is subject to a bureaucracy quite extensive for such a small area. The Hong Kong Broadcast Authority, consisting of a board of three government officials and nine non-government members, oversees the operations of the stations. Ultimate control, which includes the government's power to censor programming, is under the Department of Culture, Recreation, and Broadcasting (DCRB), a cabinet-level department that has two branches to implement its supervision: a Television and Entertainment Licensing Authority and Radio Television Hong Kong. It is expected that all media in Hong Kong will become government operated in 1997.

Korea

Following its liberation from Japan, Korea was divided into two countries, North Korea and South Korea. Although at one time privately owned stations were permitted, in both countries the government eventually took control of broadcasting. In South Korea, radio and television became public-owned and -operated media. In North Korea, both the television and radio services are state-run.

Taiwan

Taiwan has both a government-run system and a private system similar to that in the United States, with the private stations competing with each other for audiences and advertising money. However, the legacy of the Kuomintang military government has continued the pressure by the government on the media to ensure acceptable content.

Southeast Asia

Thailand has a system similar to that of Taiwan, with government and private networks, and with all stations careful not to alienate the party in power. The state keeps a close eye on all key personnel. Control over the commercial stations ensures the government of its share of advertising revenues at these stations. Broadcasting programming, content, technical and business operations, operating hours, and advertising are all under the control of the Thai Broadcasting Directing Board, which reports directly to the Office of the Prime Minister. The military controls most—literally hundreds—of the radio stations, which operate as commercial stations.

In Singapore, the government's Department of Broadcasting controls all radio and television. In Indonesia, the government's control includes revocation of a station's license and imprisonment of its personnel for violations of the Press Act, which gives the government authority to "guide" all media in their presentation of news and information and other programming. That Indonesia consists of 14,000 islands makes instant electronic communications more critical than in most other countries. While there is a dual system of commercial stations and the state-operated Radio Republik Indonesia and Television Republik Indonesia, all media are under the ultimate control of the Ministry of Information. A non-government pay-TV network happens to be operated by Indonesia's royal family.

In Burma, still under military dictatorship as this is written, the electronic media are under total state control. However, Burma's attempts to restrict its people to carefully selected and very limited information have resulted in a greatly underfunded radio and television system, and the electronic media are not as significant in Burma as in many other countries. In Malaysia the Ministry of Information controls all broadcasting, state-owned and private, through a Director-General of Broadcasting. The private stations are carefully staffed with government-approved personnel to guarantee approved content. Sri Lanka is an anomaly in that its government telecommunications system has been gradually undergoing a change to private ownership. However, its policies are mainstream in that the government maintains sufficient control to ensure content supportive of itself.

India is also an anomaly. It inherited not only Britain's democratic tradition when it achieved its independence in 1947, but British bureaucracy as well. Broadcasting under British rule was tightly controlled by the colonial government. Both radio and television are now public monopolies under the Indian Ministry for Information and Broadcasting. All programming must be approved by a government-dominated committee.

In Pakistan, at one time merged with India, broadcasting follows the same pattern: a state monopoly under the Ministry of Information and Broadcasting. All stations must adhere to programming guidelines established by the ministry. Telecommunications development seems to be slower in Pakistan than in many other developing countries in Asia, principally due to the lack of funds. All Pakistan Air operates the radio system, and the Pakistan Television Corporation operates television.

Bangladesh is the third country that once made up greater India; it was later part of Pakistan, and ultimately became an independent nation. It is the poorest of those countries and one of the poorest in the world, and its economic conditions preclude the effective development of the electronic media. Its radio and TV stations, plus its satellite receiving service, are under government control.

The importance of media to governments in their maintenance of power over the people remains unchanged even after revolutions where the people have overcome oppressive rulers. Sri Lanka is an example of this (see Appendix, p. 86). During the time it was Ceylon, its government maintained strong supervision of broadcasting operations and content. The revolution of 1971 that changed the nation into Sri Lanka did not, however, revolutionize the media. Sri Lanka continues to use the media as a tool for strengthening the party in power. Not only are the media careful to avoid any criticism of the government, but they take pains to praise its operations and agencies, such as the armed forces and the police.

OCEANIA

The Pacific Basin is a mixture of first- and third-world countries. Combinations of economic status, colonial past, current political stability or instability, strategic location and alliances, and topography determine the nature of the telecommunications systems in this region. For countries such as Australia and New Zealand, both their economic status and their memberships in the British Commonwealth have resulted in communications systems similar to that of the United Kingdom, including BBC-type government networks. Both belong to the Commonwealth Broadcasting Corporation, which was established in 1974 as a formal successor to an association of Commonwealth countries that was initially developed in World War II for more effective use of communications in fighting the Axis powers.

Australia

The Australia Broadcasting Tribunal is the nation's regulatory office, particularly in relation to privately owned stations. Established in 1976, it issues, renews, and may revoke licenses; enforces monopoly, duopoly, and cross-ownership rules; establishes guidelines for advertising; and administers rules similar to the United States' political equal time rule and former fairness doctrine. The government public corporation similar to the BBC, the Australia Broadcasting Corporation (ABC), has a board of directors that is directly responsible to the Australian Parliament and the Minister of Communications.

In addition to having stations serving the metropolitan areas, Australia, by nature of its sparsely populated mid-continent and Outback, has set up special services to reach regional stations. The vast distances have prompted Australia to utilize satellites and computerized broadcasting to reach more of its population. The School of the Air in Alice Springs, famous for its formal education through two-way radio to isolated young and old people, has in recent years added television transmission.

In recent years Australian broadcasting has been plagued by indecisiveness in government funding and regulation. From 1987 to 1993 the nation had eight different communications ministers and, in an attempt to catch up with the new technologies, passed a new Broadcasting Services Act in 1992. However, the act is so unclear that it is expected that the courts will have to decide on the legitimacy of some of the new technologies, such as microwave distribution, before they are adopted. However, one new development in 1993 was the establishment of pay-TV; Australia was the only remaining English-speaking nation in the world without subscription television.

New Zealand

New Zealand has a similar system, with radio and television network systems under the New Zealand Broadcasting Council. Both countries have private broadcast stations, with Rupert Murdoch, a former Australian who became an American citizen, thus qualifying for ownership of U.S. broadcast properties, still retaining control of much of the private system in Australia and owning media in New Zealand, as well.

Similar to the ABC is the New Zealand Broadcasting Corporation, which runs the national radio and television services. While there is rarely any attempt by the government to control the content of programs produced in New Zealand, the government does screen all foreign programming coming into the country. The government of New Zealand, which in the early 1990s began to sell off to private—including foreign—interests much of its public utilities and other holdings, is in danger of spawning a completely privatized system that has little responsibility to the country and provides little or no access for the people of New Zealand.

The Islands

Some of the islands in the South Pacific are independent nations and operate their own systems or, where they do not have the resources to do so, allow private operators to own and operate the stations. Some have mixed private and public systems. Few of the islands have television, in part because of the cost, but also because the mountainous configurations of some islands and the widely dispersed multi-island chains of some of the nations make it technologically difficult for TV to function. As noted earlier, some islands have cable distribution facilities. Radio is the principal medium on most islands. Some islands do a large amount of their own programming. Most use extensive pickups from stations in Australia and New Zealand. Nations that are still territories of larger powers import programs supplied by those powers.

By and large these governments do not seem to overtly exercise the kinds of total content controls that some developing nations in other areas of the world do. However, where the government does not operate the stations it may need to subsidize them, and the private operators are therefore sensitive to the needs and desires of the government and provide programming that strengthens the cultural and political unity of the country.

Some islands are directly dependent upon professionals from Australia or New Zealand to operate their systems; the control and operation models of those two countries are thereby making their way into the other nations of Oceania.

The lack of development of media systems in much of the area is due not only to lack of financial resources and geography, but also the weather. The humidity and salt-filled atmosphere result in rather rapid deterioration of equipment from rust and erosion.

Appendix to Chapter 3

Television's Last Frontier: Sri Lanka

Indra de Silva

(Former Head of Audience Survey and Research, Sri Lanka Television Corporation)

Situated about 22 miles from the southern tip of India, Sri Lanka is an island nation with a population of nearly 18 million. The daily press of Sri Lanka dates back to 1830 and has a circulation of more than half a million. More than one million copies of weekly newsmagazines are sold to a public whose literacy rate (87 percent) ranks at the very top among the Third World countries. Introduced in 1925, radio penetration is near universal and has more than 12 million regular listeners. Before the introduction of TV in 1979, 357 movie theaters sold 75 million tickets annually. Since its inception radio has been under government control and has played a partisan role in the national political arena, overtly supporting the ruling party. Newspapers, on the other hand, operated independently with occasional direct and indirect government censorship.

As in most Third World countries, TV came to the capital city of Sri Lanka (Colombo and its vicinity) first as a private domestic venture, called the Independent Television Network (ITN), in April 1979. Initially, ITN provided about four to five hours of canned programs mainly of U.S. origin. For example, a typical day's program began with a children's program such as "Sesame Street" or "3-2-1 Contact," followed by such programs as "Bionic Woman," "Kojak," "Mannix," "Mission Impossible," and "Petrocelli." Within its first months of operation, however, it was taken over by the government and placed under the direct control of a government appointed Competent Authority.

During the first year of operation there were only 2,810 licensed TV sets. As in many countries, one needs to pay an annual license fee to possess a TV or radio receiver in Sri Lanka. The funds collected through licensing, together with advertising revenues, are used to finance production and program procurement. Even with its limited coverage and few hours of operation, television became such a popular medium that within a year the official count of TV sets surpassed the 40,000 mark. By the time national coverage began in 1982 there were nearly 60,000 licensed TV sets in operation in Sri Lanka.

In February 1982 a second channel covering virtually the entire island was inaugurated, courtesy of Japanese aid. The Japanese government provided the state-of-the-art production and transmission equipment. As a result, at the end of the first year of the national coverage the number of households with a TV set increased nearly 300 percent. With its own production facilities and mobile production units, the daily program fare included local news, documentaries and soap operas produced in vernacular language. Yet, nearly 50 percent of the programs were of foreign origin and the dominance of foreign fare was eminent in the prime-time entertainment schedule. However, unlike the case of most Third World nations, Sri Lankans developed a taste for the locally made programs although their production values came nowhere near the slick Hollywood programs. Audience research repeatedly found a growing appetite for locally made programs, primarily in the rural areas where nearly 85 percent of the people live. Advertisers quickly responded to the popular taste by sponsoring locally made programs and within the first three years of television there were nearly 10 independent production studios making programs for national TV under the auspices of major advertisers.

The production of news, however, has been kept under the tight control of the government, primarily covering governmental activities and related stories. A content analysis of television news during the third year of operation found that nearly 60 percent of the news coverage dealt with government ceremonies involving the ministers and the members of the parliament of the ruling party. The same analysis found that absolutely no coverage was given to the opposition political parties. Moreover, only about one-fourth of the news stories consisted of spontaneous items.

The global trend towards the privatization of media systems in the eighties have also had an impact on Sri Lanka's television. Over the last seven years, for example, the government of Sri Lanka allowed four private companies to start television transmission, bringing the total number of stations to six. It is well known that the new licensees have close contacts with the ruling party. The new stations, however, were not granted the rights to broadcast news.

With the change of government in August 1994, there appears to be some policy changes with regard to television broadcasting in Sri Lanka. For example, in an interview given to one of the leading newspapers, the new minister in charge of broadcasting indicated the government's desire to allow independent stations to produce news—a major change in a country where electronic media is tightly under government control. The implementation of such promises, however, is yet to be seen.

4 Financing Global Electronic Media

As stated earlier, the sophistication of telecommunications in a given region or country is dependent on the economics of that area or nation. The "have" nations have well-developed systems, while the "have-not" nations are, with few exceptions, barely able to sustain their systems, even though they give them priority for political purposes.

Governments with some degree of affluence are able to support their telecommunications systems by subsidy, either in toto or in part. Some of these systems are state owned and operated; others are public corporations. In some instances where lack of funds has prompted the government to authorize private stations as the means by which to have any viable broadcasting system, the government provides some subsidies to keep the private operations afloat. In all cases, more and more countries are becoming dependent on greater or lesser degrees of commercial advertising for economic support. In some poorer countries, where systems are privatized with little or no restrictions on commercial advertising, the economy does not produce enough commercial revenue for these stations, and subsidies are necessary. Some governments place such value on the media for political, cultural, or educational purposes, that they provide subsidies to private stations as a matter of course.

In some countries, such as the United States and others where systems are totally or largely privatized, and where commercial advertising provides sufficient base for support and even profit, the government provides no funding whatever to broadcasting, with the exception in some cases, as in the United States, for noncommercial and/or public systems.

While commercial advertising in television and radio are frequently—and erroneously—associated solely or principally with capitalist and industrial countries where consumer goods and monies to purchase them are plentiful, this chapter shows that advertising has already spread to virtually every broadcast system in the world, including third-world and socialist nations. More significant, perhaps, is that the production and origination of commercials are spreading to many countries, too, as well as proliferating in volume.

Illustrative of the increasing importance of commercial advertising is the annual London International Advertising Awards ceremony honoring what the judges consider the best advertising in television, radio, print, and cinema. Keep in mind that in many countries movie houses show commercials, although unlike the ads on U.S. television, they are rarely permitted to be shown other than before and after the complete film presentation. The London International Advertising Awards have been given since 1968 and include political as well as product commercials. The current Awards videotape of winning commercials over the years (available to educational institutions) includes TV spots from such diverse sources as Argentina, Australia, Brazil, Canada, France, Germany, Japan, Mexico, New Zealand, Scandinavia, South Africa, Spain, and the United Kingdom.

In addition to advertising as one source of revenue, most developed countries raise funds for their non-privatized systems through a user fee on receivers. In a relatively few instances fees are charged to station operators, or taxes are placed on equipment used in transmitting or receiving signals.

NORTH AMERICA
United States

The United States is the principal private system in the world supported by advertising, but it did not begin that way. Although the first station officially went on the air in 1920, the first commercial was not aired until 1922, and then, because of tie-in contracts by AT&T, the principal broadcast equipment manufacturer, for a sharing of commercial revenue, it wasn't until almost a decade passed before commercialization was dominating the field. All networks and broadcast stations in the United States sell commercial advertising time, with the cost per time segment dependent on the rating points the advertiser will get. That's why ratings, rather than any intrinsic quality of a program, have become the determinant of all programming in the United States. Even the noncommercial or public stations seek "underwriters" and in many cases obtain such support with fee schedules similar to those of commercial rate cards. About 15 percent of public broadcasting's budget, considerably less than most Americans believe, comes from the general tax funds appropriated yearly by the federal government.

In the United States, as in other countries, cable and VCRs have cut into the ratings of broadcast networks and stations and, thus, the income of broadcasting. While cable networks also depend on advertising, most cable networks arrange for a per-subscriber fee from the cable companies that carry their programming. Some so-called "premium" cable channels charge the subscriber a fee directly, with a percentage of that fee going to the local cable company.

Canada

Although private commercial-based broadcasting is part of its system, Canada spends about 1 percent of its total annual government budget on the Canadian Broadcasting Company (CBC)—a comparatively large amount—considering it important for the growth of national culture and education. Even so, some 20 percent of CBC's annual budget comes from advertising. Canada also bolsters its private stations through special tax breaks and through special production funds. As in the United States, Canada's broadcast stations have been hard hit in recent years by a very high penetration of cable. While government funds support CBC, the private stations in Canada are dependent upon commercial advertising for their income.

Both the United States and Canada provide programming to foreign markets as one source of income. The United States, however, far exceeds Canada's foreign distribution. With English having become the unofficial second language throughout much of the world, English-language programs have become extremely popular globally. Further, U.S. sitcoms and adventure series have gained popularity in international syndication to the same extent that they did when in first-run in the United States.

Central America

As noted earlier, advertising is a high-income enterprise in Central America, and in some countries as much as 25 percent of total air time is devoted to commercials. A relatively small percentage of financial support, even in countries with government or public systems, comes from user license fees. Some systems that do not earn enough through advertising are subsidized by their governments from general tax revenues. Individual cable systems obtain their revenues principally through subscription fees.

Mexico

Mexico's broadcasting system, which exemplifies most of the others in Central America, is largely dependent on advertising and tax monies. Mexico's dual system of government and private stations moved toward total private control in 1993 when the government sold its two television networks to a private group. Such sales netted the government more than $22 billion dollars in the late 1980s and early 1990s. This has changed the annual funding from a substantial amount of tax dollars to a virtual total advertiser-supported private system. The same approach to raising funds for economically strapped governments seems to be becoming a trend in Latin America, both north and south.

In addition, one of Mexico's sources of funding is similar to the approach used by a number of countries throughout the world, especially those with third world economies. Its first satellite-receive-and-transmit system was financed largely with loans from wealthier countries. This enabled Mexico not only to send its telecommunications signals throughout its own country for the first time, but to establish communications exchange with other countries in the Americas and in Europe. Some developed countries that have actual or potential economic investments in another country find a quid pro quo situation in financing telecommunications systems in the poorer countries, mainly through opening up new trade markets.

Haiti exemplifies the system in smaller countries where there are both private and government systems: the private stations are supported by advertising and, in a few cases, as with Haiti, by viewer and listener subscriptions; and the public or government stations are supported by the state.

SOUTH AMERICA

A recurring factor in financing telecommunications is the political philosophy of the party in power. In a number of countries, such as Argentina and Peru, broadcasting has changed from government to private control and back again, resulting in both more and less advertising support for the media. In some countries, such as Colombia, advertising has been used to fund noncommercial educational programming. In other countries, such as Chile, the funding for some stations comes from their university operators, which in turn may use, in some instances, commercial advertising to provide them some measure of financial support.

In most countries the media, both radio and television, began as government monopolies. Gradually, privatization occurred. In some instances, where financing has become difficult, foreign capital has been called in, providing the investors with a share of the station or system and, of course, a share of profits. As advertising money has become more available in various countries, so have the number of broadcast hours for commercial sponsorship. Certain multinational companies, which for decades exploited Central and South American economies before they were ousted through political revolutions, have in recent years found another method of making money in these countries: selling their products, principally by promoting them through television and radio advertising.

Brazil

Brazil is a prime example of this source of financial support. TV advertising, enormous amounts from multinational companies, surpassed all other advertising means in the 1970s and since then Brazil has become one of the top five countries in the world in advertising expenditures and income. The availability of advertising money has resulted in more than 25 percent of Brazil's air time being devoted to commercials. Most of the stations in Brazil are privately owned.

When advertising became a significant factor in Peru, the number of privately owned stations expanded rapidly. Income for Peru's TV stations comes mainly from advertising. Radio in Peru has additional sources, including taxes on imported programs and fees from private stations, which are used to support the public/government broadcasting system.

Venezuela

In Venezuela private, advertising-supported stations not only grew, but did so by deliberately following the U.S. model, with the three major U.S. networks holding financial and operating stakes in the three major Venezuela networks. An interesting bit of irony is that in some countries where the government sold off or allowed its stations to go to private interests, those governments from time to time are obliged tp purchase commercial time to get their messages out to the public. Among the other countries relying principally on heavy advertising for funding broadcast operations, including public systems, are Argentina and Columbia. A few countries, such as Venezuela, use some federal tax monies to support broadcasting. Broadcasters in several countries add to their income by producing soap operas for distribution to smaller countries in Latin America that have limited production resources.

EUROPE

No matter what the system of government, private stations are increasing in Europe, following the trend in other parts of the world. In part this is due to the political changes in many areas of Europe in the early 1990s. A number of socialist states, where the media were controlled and supported by the government, moved to instant market economies, where the opportunities to make quick and fast money included the advertising revenue possible from private commercial stations. This is true not only in eastern Europe. In the United Kingdom, for example, the eighty commercial radio stations, allocated one to each market area throughout the British Isles, were expanding in 1994 to an additional 220 stations. In France commercial radio was authorized in 1982 and in two years more than 1400 stations were on the air. Many of these stations are supported by capital investments from foreign countries, in large part from the United States (Figures 4-1 and 4-2).

Financial support of telecommunications in Europe follows the usual patterns: commercial advertising, user license fees, government appropriations from general or specified funds, support by operating institutions such as educational, religious, political, and labor groups, and foreign sales of programming.

The media in all countries get at least some funding from advertising. In some countries the role of government prevents, or at least inhibits, stations from creating programs that appeal to the LCD in order to get the highest ratings and, therefore, the highest ad revenues. Some countries have established offices to keep a watch on advertising and to try to prevent excesses, as for example, the Advertising Control Division of the IBA in the United Kingdom and STER in Holland. Further, most countries have attempted to maintain the integrity of programs by mandating one or more commercial-free channels and by limiting not only the amount of time permitted for ads, but by requiring their placement in blocks that will not disturb either individual programs or program

Growth of Commercial Radio

Year	No. of Services on air at year end*	Approx. coverage of UK Pop. ILR only (incl. National) %
1973	3	25.4
1974	9	40.2
1975	16	56.5
1976	19	60.5
1977	19	60.5
1978	19	60.5
1979	19	60.5
1980	26	67.2
1981	33	73.0
1982	38	76.8
1983	42	81.0
1984	48	86.4
1985	49	87.1
1986	49	87.1
1987	50	89.5
1988	60	90.0
1989	76	92.0
1990	106	93.0
1991	106 (1)	93.0
1992	119 (2)	95.0 (100.0)
1993	130 (3) — Predicted	96.0 (100.0)

* Including split services and national services, the numbers of the latter are indicated by figures in brackets.

FIGURE 4-1
Commercial radio growth shows a significant rise in two decades. Courtesy Independent Radio.

schedules as a whole (Figures 4-3, 4-4, and 4-5).

United Kingdom

While the BBC in the United Kingdom is dependent on government funding through the collection of license fees on receivers, the Independent Television (ITV) stations must depend on advertising for their budgets. A unique arrangement exists for Channel 4, an independent creative TV network on the air since 1982. Previously subsidized by the other ITV stations, in 1993 it became self-subsistent, provided it garnered a 14 percent share of total TV advertising revenues; anything below that would be made up by the ITV stations, and anything above that would be shared with the ITV stations.

France

France tried an interesting incentive to prevent commercial needs from debilitating program quality: a "point system" to award funds to stations that produce quality programs. However, the insistence of the public on popular LCD-type programming made the point system a failure. Some countries have begun to follow France's approach in permitting sponsorship of—and therefore commercials within—individual programs.

France's telecommunications finances are broadly based. The public TV stations are supported by license fees on TV sets and foreign sales of programs, as well as by advertising. The private TV stations are principally supported by commercials. Canal Plus, the station that has garnered the greatest audience attention, includes subscription fees from viewers in its income.

Radio in France is somewhat different from that in other European countries. There are no license fees on radio sets. No advertising is permitted on the government's national stations, but the government stations in the border areas are allowed some advertising. The government maintains a strong hand in the development of even private systems, for example providing the principal fiscal support for the emerging nationwide cable system, with established cable systems obtaining their principal support from subscription fees.

Scandinavia

While all private systems in Europe have commercial advertising, not all public systems do. Scandinavian countries were among the last to hold out against commercialization, but in the early 1990s, as telecommunications costs

Audience Share by Radio Service

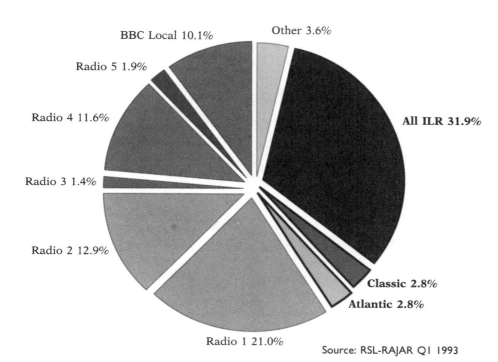

Source: RSL-RAJAR QI 1993

FIGURE 4-2
Commercial radio stations draw the largest audience in the United Kingdom. Courtesy Independent Radio.

More People Listen to Commercial Radio

Due to differences in methodology direct comparisons cannot be made between RAJAR and JICRAR.

However, RAJAR generally captures lower levels of *measured* listening than JICRAR and thus the increases in listening to commercial radio indicated by the figures below are all the more significant.

	Weekly Reach	Total Hours	% Share of Listening	Coverage of UK %
1988 (JICRAR)	17667	232253	30.8	90.0
1991 (JICRAR)	22593	313947	35.6	93.0
1993 (RAJAR)	25380	324434	37.4	100.0

Note: figures relate to listening within the coverage area of commercial radio.
Source: JICRAR: 1988 & QI, 1991/RAJAR: QI 1993

FIGURE 4-3
Independent AM radio service in the United Kingdom's Midlands. Courtesy Independent Radio.

its budget needs. Danish radio, however, still has relatively little advertising, and is principally dependent on license fees. The country's growing cable system relies on advertising. The competition in the early 1990s between Denmark's two TV networks resulted in a ratings war, in which each station counterprogrammed against the other, similar to the practice in the U.S., in order to obtain higher ratings and consequently higher commercial income.

In Norway the government stations are supported by income from license fees and, as necessary, subsidies from general tax revenues. However, satellite and cable television operations obtain their funding from advertising. In Finland, receiver license fees and government subsidies account for the bulk of the budget of YLE, the government broadcasting system. YLE is also supported in part by fees paid by private companies for rental of air time on the YLE system. A new commercial TV network that began operations in the early 1990s has some advertising, but thus far it has been limited. Finland's local radio stations are permitted some advertising, but the national radio network still has none. The commercial TV network obtains its capital from hundreds of investors, including the federal government.

Sveriges Radio, Sweden's system, is a joint stock company monopoly, with private investors, the press, and other entities as partners. All of its funding comes from license fees, although the government provides additional general tax revenue subsidies for news and information programs. However, as this is being written, Sweden is venturing into the realm of advertiser-supported electronic media.

While all countries in western Europe impose user license fees for TV sets, not all do for radio sets. France, Luxembourg, and Norway are the prime exceptions. User fees are usually collected by the government office responsible for communications regulation—a Ministry of Posts and Communications, for

grew and the economies of many European countries fell into the depression that began in the United States several years earlier, the need for additional funding to maintain telecommunications also grew (Figure 4-6). Reluctantly, but satisfactorily from an income point of view, commercial advertising gradually made its way into Scandinavian systems. For example, Denmark introduced commercials on its TV system when its first commercial station, TV2, found that 70 percent of its budget could be provided by advertising, requiring license fee funds for only 30 percent of

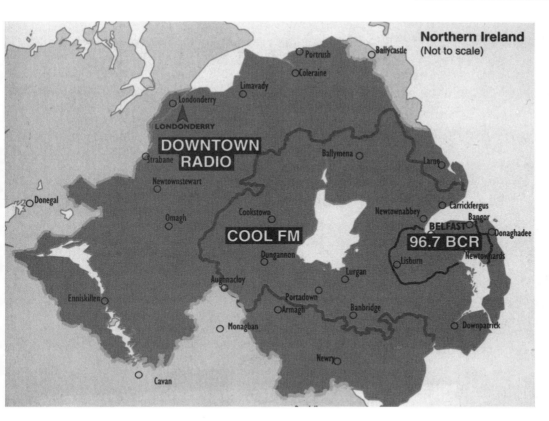

FIGURE 4-4
Independent FM radio service in Northern Ireland. Courtesy Independent Radio.

example, with the legislative body, such as a parliament, usually authorizing the distribution of the collected fees to systems and/or stations.

Because user fees are generally unpopular, some countries try to avoid any raises in the fees by adding annual appropriations from general tax receipts to the funding for telecommunications systems. Broadcasters prefer a fee-only approach, provided it gives sufficient funding, because dependence on annual appropriations also entails parliamentary debate about the system and includes concerns and pressures regarding programming and operations.

Portugal

Some countries, such as Portugal, have different combinations of support for different services. Portugal's public system depends on user license fees, advertising, and government appropriations.

Private stations, which began to appear in the early 1990s, resulted in greater increases in advertising and financial stability for those stations. Radio in Portugal receives its funding mainly from license fees and government subsidies; television, however, gets its major support from a combination of license fees and advertising.

Switzerland

Switzerland takes somewhat the same approach, with television receiving its support from license fees and some restricted advertising, but radio is permitted no advertising and is dependent on license fees. Switzerland has a dual system. Government-operated stations are funded through advertising, license fees, and some government appropriations. Private stations must rely on advertising. Switzerland also has a pay

FIGURE 4-5
Independent FM
radio in Scotland.
Courtesy Indepen-
dent Radio.

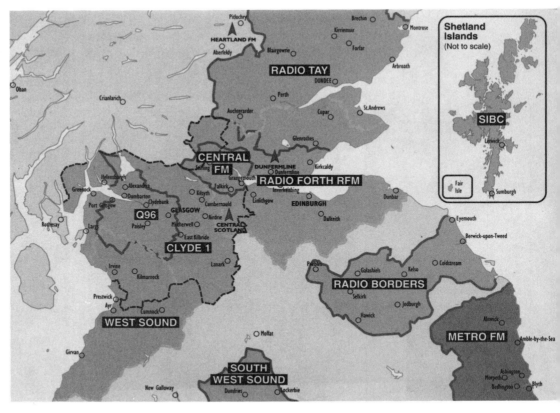

television network which is supported by subscriber fees.

Institutional support is usually present on the local, as opposed to national, level in a number of European countries. A given organization might well afford to operate a local station but would be unable to afford the much greater funding necessary for regional or national stations. In some cases private industry provides institutional support. As the electronic media grew in various countries in Europe, advertising revenue for newspapers was lost to television and radio. When this happened in the United States, newspapers began to buy up stations, a practice which was halted in the 1970s with the FCC's cross-ownership rule. Such acquisition is happening in Europe and in other parts of the world where there are no legal bars to media monopolies. In Germany, for example, funding comes principally from license fees on receivers. However, at the regional public stations there is some advertising to add to the financial sup-

port from these fees allocated by the Conference of German Broadcasting Stations. In addition, the burgeoning private telecommunications system, mainly cable owned and operated on a commercial basis, has seen a growth in private investors, especially newspapers. Cable's finances are also added to by subscriber fees, similar to the process in the United States.

A number of countries in Europe are selling programs to foreign countries. Leading exporters are the United Kingdom all over the world, and France, Germany, and Spain to nations where their languages are spoken, which results in income to the exporters for both public and private systems.

The financing structures are, of course, different for the public or government systems and the private systems, and in many cases, different for radio and for television. All the private television systems in western Europe depend on advertising, with two countries, Ireland and Iceland, also permitting private stations to elicit subscriber fees. The public or government television systems vary in sources of funding. While all but Luxembourg, which has no public system, have user license fees, France, Germany, Ireland, and the Netherlands rely as well on advertising. Germany, in addition, provides governmental appropriations, permits the sale of program guides, allows membership fees, and has a substantial foreign sale of its programs. France also has a large foreign sale. Finland supports its public network, YLE, in part from fees paid by the private network, MTV. For radio, France, Ireland, Luxembourg and Finland rely primarily on advertising, while Belgium, Denmark, Germany, Iceland, the Netherlands, and the United Kingdom have both advertising and license fees.

As costs for telecommunications operations increase, more and more countries are adding advertising to the user fee and/or subsidy approach. Ireland relied primarily on annual license fees, but recently permitted a growth in advertising. The use of ads is rampant in Belgium. Austria reluctantly added advertising fairly recently, with 70 percent of station budgets coming from license fees and the rest from ads. The Netherlands has come to about the same mix with its cable system being supported by a combination of subscription fees and ads, as is done in Germany. Cable in Holland, however, is dependent on local and regional support inasmuch as the federal government is reluctant to provide any funds for its operations. RTL4, the station for Dutch viewers beamed from Luxembourg, makes its money from heavy advertising.

Germany

Germany relies mainly on receiver license fees for support of its regional systems, although a limited amount of advertising is done on some of the stations. The unification of East and West Germany added to the base of support, with fees for eastern Germans considerably less than those for westerners in

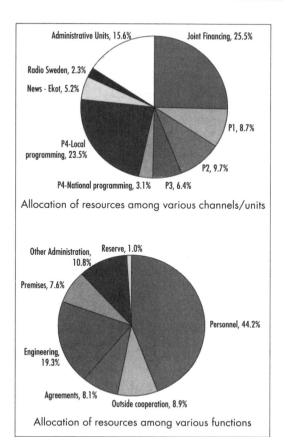

FIGURE 4-6
Swedish Broadcasting Corporation's allocation of funding. Courtesy SBC.

order to slowly integrate the economy of the east into that of the west (see Figure 4-7). However, in 1995 fees were to be equalized for all Germans, with levies on one TV and one radio set, but no additional cost for additional sets. Another source of money for German telecommunications comes from newspaper investments. As happened in the United States with the early growth of radio, newspapers in Germany are trying to join or, in fact, buy the competition, rather than fight it. License fees from private cable systems add to the government's sources of funds for public broadcast systems.

Southern Europe

Countries in southern Europe are following somewhat the same pattern, although most of these nations are not as economically sound as their northern neighbors. The rise of private stations and their popular programming has

FIGURE 4-7
Rate cards for commercial radio stations in Germany. Courtesy Antenne Bayern.

LOKALFUNK KOMBI WESTFALEN

Lokalradio kommt an.

Name der Werbekombi
LOKALFUNK KOMBI WESTFALEN

Tarifzusammenschluß der Sender:
Westmünsterland-Welle, Radio FiV, Radio Kiepenkerl, radio RST, Radio MK, radio Antenne Münster, Radio WAF, Radio Lippe Welle Hamm, Hellweg-Radio, Radio Gütersloh, Radio Westfalica, Radio Bielefeld, Radio Hochstift, Radio Herford, Radio Lippe

Name des Anbieters
AUDIO MEDIA SERVICE Produktionsgesellschaft mbH & Co. KG

Adresse
Niedernstr. 21-27, 33602 Bielefeld
Telefon (0521) 555170
Telefon (0521) 555180/81
Telefax (0521) 555152

Verkaufsleiter
Klaus Förster

Disposition
Alexandra Kahlert, Andrea Rieke

Sendegebiet
Westfalen

Techn. Reichweite
4,9 Mio. Einwohner

Tatsächl. Reichweite
1,3 Mio Hörer gestern Mo.-Fr. in NRW
310.000 Hörer/Std. Mo.-Sa. 6.00-18.00 Uhr

Zielgruppe
18 bis 49 Jahre

Gültiger Tarif
Nr. 3 vom 1.1.1994

Spotpreise

Mo.-Sa.	DM/Sek.	DM/30 Sek.
06.00-10.00	96.-	2.880.-
10.00-14.00	76.-	2.280.-
14.00-16.00	53.-	1.590.-
16.00-18.00	62.-	1.860.-
18.00-06.00	Sonder-Sek.-Preis für alle 8 Sender	

So- u. Feiertag	DM/Sek.	DM/30 Sek.
06.00-18.00	53.-	1.590.-
18.00-06.00	Sonder-Sek.-Preis für alle 8 Sender	

Mindestspotlänge: 15 Sekunden
Alle Preise zzgl. der gesetzlichen MwSt.

Kombis

Rabatte

ab	1000 Sek.	3,0%
ab	2000 Sek.	4,0%
ab	3000 Sek.	6,0%
ab	4000 Sek.	7,5%
ab	5000 Sek.	10,0%
ab	7500 Sek.	15,0%
ab	10.000 Sek.	20,0%

Buchungsschluß
7 Arbeitstage vor dem Sendetermin, kürzere Termine nach Abstimmung

Sendeunterlagen
Tonband 38,1 cm/Sek. in Studioqualität, 1/4 Zoll. Lieferanschrift: AUDIO MEDIA SERVICE, Adresse s.o.

---

seriously affected the viability of government and public stations, with their higher quality and increasingly less-watched or listened-to programming. This has put a serious financial strain on public systems. Austria's state-run system, for example, was considering in the early 1990s the possibility of selling off a part, but not quite a controlling, interest to private organizations. Its funding from license fees and advertising still leaves it short of the financing it needs. About two-thirds of ORF's budget comes from receiver license fees, about 30 percent from advertising, with government subsidies expected to make up shortfalls.

Greece

In Greece television is funded by a combination of advertising, a fee added onto electric bills, and government subsidies, all of which are not sufficient to keep the stations in the black. Radio financing is similar, with local stations receiving funding, as well, from local governments. Like other governments, Greece

in the late 1980s saw little choice but to turn to the introduction of privately owned stations.

Italy, similarly, once a government controlled system of radio and television, now operates as a dual public and private system, with revenues principally from advertising and user fees. Privatization has resulted in huge increases in advertising, and both the public and private stations have taken advantage of this. Portugal's public system depends on user license fees, advertising, and government appropriations. Private stations that began to appear in the early 1990s prompted great increases in advertising and financial stability for these stations.

Spain

At one time in Spain the government television monopoly made enough money from advertising to support not only TV, but, in part, the public radio system, too. The private radio system supported itself. With the decentralization of the system in Spain, individual communities

FIGURE 4-7
continued

Hörfunk/Nielsen IIIb

HIT HOUSE
RADIO 107,7 STUTTGART
Mo.-So. 12⁰⁰ - 20⁰⁰   Sa. 12⁰⁰ - 5⁰⁰
THEODOR HEUSS STR. 24
70174 STUTTGART

Senderkennung
Hit-House Radio 107,7 Stuttgart

Name des Anbieters
Michael Bernthaler

Adresse
Theodor-Heuss-Str. 24
70174 Stuttgart
Telefon (0711) 2238081
Telefax (0711) 2238083

Geschäftsführer
Michael Bernthaler

Anzahl der Mitarbeiter
7

Sendebetrieb seit
27.04.1989

Marketingleiter
Bernd Gehrung

Werbezeitenverkauf
Bernd Gehrung, Jens Nill

Senderstandorte
Terrestrisch: UKW 107,7 MHz
Im Kabel: UKW 107,2 MHz (nähere Auskunft
im Sender)

Sendegebiet
Region Stuttgart - LK Böblingen - LK Esslingen
- LK Ludwigsburg - Rems-Murr

Repräsentativerhebung
Infratest Studio 1993

Techn. Reichweite
ca. 1,7 Mio.

Tatsächl. Reichweite
Hörer gestern (Mo-Fr)   71.000**
Stammhörer   77.000*
Gelegenheitshörer   165.000*
weitester Hörerkreis   245.000*
Bekanntheitsgrad   608.000*
Werte gültig für Gesamtfrequenz incl. Splitting-
partner
* Mo-Fr Sendegebiet
**Mo-Fr Baden-Württemberg

Programmkonzept
Hit-Dance Radio, Chart Shows, Lokaler + Inter-
nationaler Klatsch und Tratsch, Network
Charts, Hollywood Affairs, Star Interviews, Live
DJ Mixes, Sport News, Trendorientierte Musik,
Personality Shows

Format/Musikfarbe
European Hit + Dance Radio

Zielgruppe
14 bis 35 Jahre

Sendezeit
Täglich 17.00-20.00 Uhr / Samstags 20.00-
05.00 Uhr

Gültiger Tarif
01.10.1993

Spotpreise

Mo-Fr	17.00-20.00 Uhr	8.- DM
Sa	20.00-24.00 Uhr	6.- DM
So	00.00-05.00 Uhr	6.- DM
	17.00-20.00 Uhr	6.- DM

Rabatte
ab

500 Sek.	2%
750 Sek.	3%
1.250 Sek.	5%
2.500 Sek.	8%
5.000 Sek.	10%
10.000 Sek.	16%

Mitglied in Werbe-Kombi
Funk-Network Baden-Württemberg

Sonderwerbeformen
Sponsoring - On Air/Off Air - Partys + Live Ver-
anstaltungen und Job-Telefon

Buchungsschluß
5 Arbeitstage vor Sendung, kurzfristige Buchun-
gen sind nach Absprache möglich

Sendeunterlagen
Tonband 38,1 cm/Sek. in Studioqualität, Car-
tridges, DAT-Tapes, mit Gema-Angaben. Unter-
lagen sind 5 Tage vor Sendung zu liefern an
Hit-House, Adresse s.o.

---

and provinces began to operate and fund their own stations, partially through advertising and partially through regional government subsidies. By the 1990s government subsidies began to abate and the local stations became dependent on advertising. The private TV stations in Spain rely on advertising and on subscriber fees. The private networks provide some of the funding for the public networks by paying fees to the government for use of the government's transmitters and translators. Private commercial stations operate much as do their U.S. counterparts, relying heavily on ratings for their ad-driven financial support. Because of the strong competition, also similar to that in the United States, the fees private stations pay to the government cut heavily into their profit base.

Italy

Italy's state systems rely principally on receiver subscription fees, with about 50 percent of that income going to television and about 35 percent to radio. Advertising provides some income to the public stations and virtually all the income to private stations. However, commercials are restricted to seven minutes per hour for public stations and twelve minutes per hour for private stations. Radio earns only about a third of the advertising revenue, but despite the larger number of radio stations the total costs of operation are less than that of television. An increasing source of funding in Italy is pay-TV, with intense competition in a growing market.

Several of the smaller countries in Europe are typical of the rest of the region. Gibraltar's system is run by a commercial organization, which is funded by advertising, license fees, and government subsidies. However, it is too small to do much of its own programming, and much of its budget goes to the importation of foreign signals. Malta's state-run system, which also imports much of its programming, is also moving toward some degree of advertising and private participation. Monaco has a public and a private network, with the public system relying mainly on government funding and the private system on advertising.

Eastern Europe

Eastern Europe should be an anomaly, given its recent political changes. But in fact it is simply a reflection of the privatization taking place in the rest of Europe, albeit for different immediate reasons: the drastic changes in economic structures and policies. As in western Europe, eastern European countries use three principal sources for funding: advertising, user fees, and government appropriations from general revenues. User fees are both old and new. Most countries still follow the model set by the former USSR, a model that has not yet changed as drastically as its political systems. At one time a major source of revenue in the USSR and in some other eastern European nations, user fees were abolished in some countries and replaced with a tax on purchases of new sets. With the changes of the early 1990s, user fees are being considered again by some countries, inasmuch as government subsidies are lessening as government control lessens. Hungary, for example, uses a combination of government subsidies and advertising, while the Czech Republic relies on a combination of license fees for public stations and advertising for the recently developed private stations.

Converts to Capitalism

Advertising is not new as a reflection of a changed economic philosophy. As early as the mid-1960s the USSR welcomed U.S. advertisers, albeit with restrictions on hard-sell approaches. Since the velvet revolutions, foreign investors have begun to play large roles in media operations in eastern Europe. As in other enterprises, foreign money was necessary for media ventures.

One of the tragic by-products of media advertising as a part of instant capitalism was the 1995 murder of Vladislav Listyev, the new head of Russian Public Television, the part public and part private company that replaced the state-run television system, Ostankino. Reports from Moscow indicated that Ostankino had generated $5 million in commercials per year for the system under its standard rate scale, plus an additional $30 million under an unofficial rate scale, the latter sum skimmed off by corrupt operators and politicians. Listyev announced he would eliminate all advertising, thereby halting the fortune in illegal rakeoffs. He was promptly killed. Uncontrolled free enterprise not only in Russia but in other eastern European countries in the mid-1990s threatened to either dominate or tear apart the structure and operations of a number of broadcasting systems.

In 1993 the Russian parliament rejected a law that would have restricted foreign investments in telecommunications to no more than 40 percent. Key stations in Russia, including large TV operations in Moscow and St. Petersburg, get substantial financing from U.S. companies. In the early 1990s other eastern European countries were passing or developing legislation to encourage private investment in private stations. The Czech Republic, for instance, opened up much of its radio spectrum to new privately owned stations. One problem in attracting foreign investors to bolster or develop any given eastern European country's telecommunications system is the uncertainty of the political situation. A number of countries had not yet, by 1996, determined a clear set of rules and regulations for communications, making it risky for investors until some economic and political stability is established.

A special source of funding for telecommunications in some eastern European countries is from foundations and from other governments that are providing aid to help the region's transition from one economy to another. The lack of usable currency has limited eastern Europe nations' importation of programs, but several have solved the problem by exchanging programs on a barter basis. One phenomenon of the instant market economy endeavors in eastern Europe is similar to a practice in many other countries attempting to add to their financing base: fees charged to foreign countries to carry one or more of their broadcast channels. In the Baltic states of Latvia, Lithuania, and Estonia, Russia's Channel 1 pays broadcast fees to these countries for carriage of its signal to the sizable Russian populations that emigrated to the Baltic countries during Soviet rule and who have remained there. Estonia, for example, is about 40 percent Russian, Latvia about 50 percent. In most situations of this kind the transmitting country wants its programming to reach a larger target audience for its commercial product advertising. But here Russia is principally interested in maintaining cultural ties and political influence over its nationals in the adjacent countries. In early 1994 another facet of instant capitalism threatened the Russian-Baltic states arrangements: instant poverty. The three Baltic countries threatened to cancel carriage of the Channel 1 signal because the Russian station had been unable to pay the agreed-upon carriage fees.

AFRICA

Africa in general follows the same financing approaches as other regions, but because of third-world poverty and the difficulty of monitoring and reaching many parts of most countries, user fees are not a viable form of financing. Advertising is growing in Africa, as is privatization—similar to other regions. But because of the substandard income in most African nations, direct government funding is necessary to keep even many private stations in operation. The highly bureaucratic nature of many African countries, coupled with highly inefficient operation in some, results in unplanned and sporadic funding that precludes long-range planning and development for the systems.

In Kenya, for example, the government provides a yearly budget to add to the advertising income of the stations. The system is dependent on government funds, thus making the stations very attentive to government policies and influence as they seek new appropriations every year. Unspent funds must be returned to the treasury. Kenya has developed a broad base of support for its telecommunications growth: licensing fees, general revenue appropriations, advertising, and funds from private investors for joint government-private telecommunications projects. All stations in Kenya are government owned.

Tanzania relies on advertising to support its broadcasting system, but imposes restrictions on commercials that may reflect on the government's political and social policies. The advertising, like the programming, must serve the cultural and political unification of the country, especially in promoting peaceful and cooperative relationships among the country's many tribes. In Zambia the same approach is used. Advertising is the key to financing the broadcast system, with rules similar to those of Tanzania, although somewhat less stringent. Zimbabwe's broadcasting system is under the Ministry of Information, Posts, and Telecommunications, and is dependent upon annual government funding, which, as in other countries, gives the government continuing strong influence. Zimbabwe's media also get some income from advertising and from license fees. In addition, it uses two financing techniques used by some other countries, as well: sale of air time to nongovernmental organizations, and loans from neighboring countries.

In some countries, where the media are under the tight control of the government, virtually all of the funding comes from the government as part of its operational budget. Namibia is an example:

its National Broadcasting Corporation gets 80 percent of its budget from the government each year and is necessarily amenable to presenting content that supports the government's policies. Zambia's system is also controlled and funded by the government.

Some countries use combinations of government funding and advertising. In Zaire, for example, all advertising revenue goes directly to the government, which then uses it for its telecommunications system. Liberia has a combination of state funds and advertising. Ethiopia's system gets it support mainly from the government, but also has some advertising. Officially, there are license fees, but very few people pay them, in part because of the poverty in the country and in part because it is virtually impossible to enforce collection. Senegal's system also relies principally on government funding, but includes some advertising. Angola has recently permitted advertising, but financial support comes mostly from the government. On the other hand, Nigeria's system operates under a public authority and is dependent mainly on advertising for its budget. Chad has very little advertising and as an extremely poor country has little government financial resources for telecommunications. Like some other countries in the same situation, Chad has been greatly dependent on help from the African Development Bank.

South Africa is one of the few African countries that depends heavily on, and is successful in administering, user license fees. Fully half of broadcasting's budget comes from that source. The other half comes from advertising and, because it is a large country with a more advanced telecommunications system than most other African countries, it has some income from sales of programs to other nations and from its production services.

In a different part of Africa, the wealthy Gulf States are able to support their government-operated systems easily from government funds. Nevertheless, some of them do receive income from carefully controlled and screened commercials and through licensing advertisers who wish to put commercials on the air.

The Mideast

The mideast countries reflect in their telecommunications financing structures the political structures of their governments. In the Arab monarchies funding is dispensed by the government, which tightly controls and operates the systems. Saudi Arabia and Kuwait are examples of countries that operate in this manner.

Egypt, with a government-controlled but generally freer system, relies principally on government subsidies for radio. However, its Middle East radio service, which broadcasts to other Arab countries, also has advertising and some private investment. Egyptian television has some ads, mostly oriented to products and services in such areas as health and family growth that are deemed of special value to the viewers. Some of these ads are by multinational companies that have arranged with their government to sell their wares in Egypt. A large part of television income, however, comes from the sale of programs to other Arab countries. Israel is an anomaly in the mideast and reflects the practices of many European countries. Television is supported mainly by license fees on receivers and by government subsidies, while radio receives much of its income from advertising. Advertising is growing, providing relief from the vicissitudes of state subsidies.

Iran moved from a relatively flexible, albeit government-influenced, system that included private commercial stations to a totally controlled government operation to solidify its political control. In so doing it built support of broadcasting in its official long-range budgets. National Iranian Radio and Television (NIRT) augments its funds with auxiliary services such as communications training programs, cultural presentations, and projects and studies on the field of communications.

ASIA

Government funding, user license fees, and commercial advertising similarly provide the principal economic support for telecommunications systems in Asia, varying, of course, from country to country. As in Africa, the third-world poverty of many nations limits the levying of service fees. Because almost all telecommunications systems are operated by governments, a combination of government funding and commercial revenue is most often found. Commercial advertising has grown as international trade has increased. Asia is a prime area for transnational advertising as its economy has improved and its political systems have opened up (at least in some countries) to provide its populations with greater interaction with the rest of the world, including foreign markets. This accounts for a huge increase in transnational trade and concomitant advertising on media systems. By the early 1990s some 30 billion dollars was being spent on transnational advertising in Asian markets, with much more expected to come. The same political and economic changes in governmental philosophies that facilitated increasing interaction between first- and third-world economies also opened the door to privatization of stations in a number of countries, and this, too, has accounted for an increase in commercial revenues as a base of support for telecommunications systems. For example, from 1982 to 1993 advertising revenue increased tenfold in Malaysia, and the same percentage in Indonesia from 1989 to 1993. In both Hong Kong and Taiwan the increase was 500 percent from 1982–1993, and in Singapore from 1985–1993 and Thailand from 1986–1993 it was 300 percent.

Most countries use a combination of financial support mechanisms. Afghan-istan, for example, has government appropriations, commercial advertising, and a tax on the importation of receiving sets, and seeks subsidies from more affluent countries. Among the countries that have combined license fees and advertising are Sri Lanka and Singapore. Broadcasting in Pakistan obtains its funds principally from license fees on receivers and state subsidies, with minimal commercial advertising. However, there are different funding approaches for radio and for TV. Radio relies principally on ads and on taxes collected on radios that are imported into the country, while television's funding comes mainly from the license fees, state subsidies, and some advertising. Cambodia provides government subsidies and relies in part on advertising revenue, but has no license fee. Thailand is an example of a country that started with a government operated and financed system, but moved gradually to advertising as a necessary financial adjunct, and currently is moving toward private investment, although the government still remains the principal source of funds. In countries that have gone in this direction the expectation is that the telecommunications systems will eventually become privatized, either in whole or at least in part. This is true for Malaysia, too, which heretofore depended primarily on user license fees and commercial advertising. The need for an infusion of more money led to private investment and an introduction of privatization. Indonesia follows the same pattern, with a growing encouragement of infusing more money into its telecommunications system through foreign investors.

In a different part of Asia, with a publicly controlled broadcasting organization, Turkey has taken much the same approach to financing telecommunications. Turkey uses government funds, user fees, and commercial advertising to support the independent association running radio and TV. And, as in a few other countries, the stations find additional monies through public performances such as concerts by the arts groups associated with the broadcasting system, and by the sale of publications relating to radio and TV.

In some countries the economy is good enough to permit not only self-supporting operation of systems, but even growth. For example, the Philippines, while not on a par with first-world industrial nations, is nevertheless in better economic shape than some Asian countries. Its communications system is supported by license fees, advertising revenue, and a special tax on advertising itself. Because the economy is relatively stable, so are the private stations of the country. Singapore, one of the few countries in Asia that can boast an excellent economy, obtains much of its telecommunications funds through fees from broadcasters. The financial health of Singapore has enabled it to be on the cutting edge of new communications technologies, as well.

Countries that maintain full control and operation of their telecommunications systems do not necessarily follow the funding approach that one might expect: exclusively government appropriations. China, one of the countries that has never had a user fee, supported its system through government funding, but even during its strongest period of a socialist economy, in 1979, it introduced commercial advertising as a source of financial support. Advertising revenue has become an important part of financing radio and television in China, supplementing the state budget. In 1993 the state television network, CCTV, was deluged by advertisers wanting to pay about $3,500 U.S. for fifteen-second spot announcements (a minuscule fee in the United States, but a large amount at the time in China). The advertising boom in China was expected to earn CCTV over 150 million dollars in 1994.

Japan's dual system, NHK and private stations, has a dual system of financial support (Figure 4-8). NHK's budget depends principally on user license fees. However, they are collected on a voluntary basis and NHK has been reluctant to raise fees even as costs have risen. This requires, then, annual subsidies from general government funds. On the other hand, private stations are doing extremely well, as Japan's economy has resulted in high-volume, high-priced media advertising (Figure 4-9).

India is a good example of a country that, to its surprise, discovered commercials much like a child discovers a new ice cream flavor. At one time India supported its government operated system through appropriations and user fees. It abolished user fees and began to rely on advertising. Even in a faltering economy, it found that advertising revenues more than met operating costs, and it has encouraged the expansion of commercial usage in order to raise additional revenue to expand the system nationwide. This is happening in an increasing number of countries, not only in Asia, but throughout the world.

International politics play important roles in the financing and developing of some nations' communications systems. Neutral or buffer countries are wooed by the larger powers in competition with each other for world influence. Because of the extreme importance of the media in affecting public opinion, rival countries sometimes attempt to obtain a foothold in the growing telecommunications systems of key swing countries. A case in point is Nepal. In the early 1970s Nepal was interested in developing a television system that would provide information and education to its population in order to strengthen its role as a nonaligned country in Asia and obtain favors from countries seeking its support. Both China and the U.S. offered to help develop the new TV system. One of the authors of this book was asked by the U.S. State Department to meet in Nepal with key government officials to advise them on planning a system, concomitantly showing them how the U.S. approach was the most advantageous one. The Nepalese were in a position to bargain for the best deal. Similar situations—that is, funding and technical

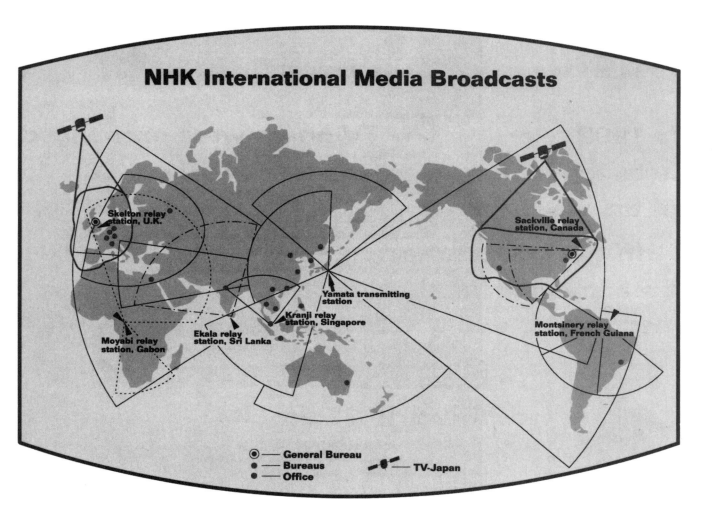

NHK International Media Broadcasts

Skelton relay
station, U.K.

Sackville relay
station, Canada

Yamata transmitting
station

Kranji relay
station, Singapore

Moyabi relay
station, Gabon

Ekala relay
station, Sri Lanka

Montsinery relay
station, French Guiana

◎ — **General Bureau**
● — **Bureaus**
● — **Office**

— **TV-Japan**

FIGURE 4-8
Japan's broadcast-
ing spans the globe.
Courtesy NHK.

and advisory assistance for telecommunications from politically motivated
nations—have made it possible for a
number of have-not countries to develop
telecommunications systems that otherwise might have lagged for an indeter-
minate time. Ultimately, the Nepalese
system grew with continuing foreign
investments, including state-of-the-art
equipment from Japan. Today, principal sources of operating funds come
from commercials, including a substantial amount from international companies' advertising.

OCEANIA

Broadcasting structures and operations
in the Pacific Basin have been influenced greatly by Australia. Financing of
systems generally follow the same approach used in Australia, which is a
combination of user license fees, government appropriations, and advertising.
The public system, the Australian Broadcasting Company, is now also funded in
part from license fees imposed on commercial stations, based on their adver-

FIGURE 4-9
Japan Broadcasting Corporation's fee fact sheet. Courtesy NHK.

NHK FACTSHEET

No. 12

Receiving Fee System

NHK is the nation's sole public broadcaster. It is financially and managerially independent of government and corporate financial support as well as programming influence. The financing mechanism was designed to instill the highest standards of impartiality in all aspects of the network's operations.

The "Receiving Fee System" makes such autonomy possible. It was mandated in Article 32 of Japan's Broadcast Law which states: "Whoever installs reception equipment capable of receiving the Corporation's broadcasts is required to conclude a contract with the Corporation (NHK) in respect to broadcast reception." Thus, everyone who can receive NHK television signals is obliged under the law to conclude a receiving contract with NHK and to pay the fee.

The following is an outline of NHK's fee system and the methods for payment.

The Broadcast Receiving Contract System

Categories of receiving contracts

NHK Receiving Contracts include an overall fee system for television, AM and FM radio. But usually contracts are either: (1) for color reception of terrestrial TV broadcasts (Color Contract); (2) for black-and-white reception of terrestrial TV broadcasts (General Contract); (3) for color reception of both satellite and terrestrial TV broadcasts (Satellite Color Contract); (4) for black-and-white reception only of both satellite and terrestrial TV broadcasts (Satellite General Contract). All these include radio services.

NHK additionally offers a special contract for satellite TV broadcasts in areas where geographical interference of reception occurs and for broadcast reception on board passenger trains or commercial transport vehicles (Special Contract).

Fee Schedules

Each fiscal year NHK submits to the Japanese Parliament a financial report and proposed budget estimating revenues and expenses expected for the coming year. Parliament reviews and analyzes the reports and proposals, and advises and /or approves. The amount of the monthly fee is calibrated with parliamentary approval of the

NHK revenue and expenditure budget proposal for each fiscal year. Current fees were set in April 1990. A color Contract is ¥1,370 and a satellite color Contract is ¥2,300 monthly with various discounts, depending on the method of payment. In 1993, receiving fees accounted for 96.5% of all NHK operating revenues.

Fee Payment Method

Receiving fees can be paid by automatic account transfer, postal payment order, or to door-to-door collectors. Eighty business centers have been established in broadcasting stations and service offices nationwide to service contracts. Banks, all post offices, certain convenience stores, and other organizations also accept NHK fee payments.

Payment Periods

There are 6 two-month payment periods annually. Discounts for advance payment are about 5% or 7.5%.

Payment Exemptions

NHK exempts various social welfare facilities, educational institutions and the needy from receiving fee payment obligations. A discount of 50% is available to people with hearing, vision or severe physical disabilities. Victims of natural disasters may also qualify for 2-month exemptions.

Current Receiving Contracts and Fee Collection

Number of contracts

The total number of contracts for NHK broadcast reception at the end of January, 1994 was 34.6 million, of which Satellite Contracts accounted for 5.7 million, an increase of 861 thousands over the previous year.

Audience Relations

Fee collection rates are at 97%, but NHK business centers have a responsibility beyond drawing up contracts and collecting fees. They facilitate communication between NHK and its audience through telephone calls, letters and

Public Relations Bureau Nippon Hoso Kyokai (Japan Broadcasting Corporation)
2-2-1 Jinnan, Shibuya-ku, Tokyo 150-01, Japan Tel: +81.3.3465-1111, Telefax: +81.3.3469-8110

April, 1994

ADVERTISING STANDARDS AUTHORITY

The Advertising Standards Authority Inc. (ASA) was formed in 1973. Its brief is to maintain advertising standards in New Zealand.

ASA's membership consists of representatives of the Advertising Agencies Association, Association of New Zealand Advertisers, Newspaper Publishers' Association, Community Newspapers Association, Magazine Publishers' Association, Independent Broadcasters' Association, Radio New Zealand, Television New Zealand, TV3 and the New Zealand Cinema Advertisers Council.

It has three main functions:

1. To introduce Codes of Practice that embody proper, truthful and generally acceptable standards of advertising.
2. To maintain and improve these standards through self-regulation by media and advertising groups.
3. To fund and resource the ongoing operations of the Advertising Standards Complaints Board.

ADVERTISING STANDARDS COMPLAINTS BOARD

The Advertising Standards Complaints Board was established by the ASA in March 1988.

The Board is the body which considers complaints on advertisements. This Board has three key roles:

1. Adjudicating in cases of advertisements where there are alleged breaches of the Codes of Practice.
2. Advising the Advertising Standards Authority (ASA) on improvements to Codes and advertising standards.
3. Reporting to ASA on any aspect of advertising which is causing concern.

A feature of the Complaints Board is that four of its eight members, including the chairperson, are public representatives with no connection to media or advertising groups. This public representation is further strengthened by allowing the chairperson, where appropriate, to exercise a casting vote.

CODES OF PRACTICE

There is an Advertising Code of Ethics which sets out the general standards for all advertising. Additionally there are particular Codes for specific categories of advertising where they are considered necessary.

Currently there are Codes for the following:

- liquor advertising
- cigarette advertising
- baldness or hairloss
- driving and petrol consumption claims
- financial advertising
- the way people are portrayed in advertising
- slimming or weight loss
- reproduction of banknote images
- farm safety
- advertisements for children
- environmental claims
- comparative advertising

The Codes, where appropriate, have been developed in consultation with industry, consumer groups, and government departments. Copies of any Codes are available on request from:

The Secretary,
Advertising Standards
Complaints Board,
P.O. Box 10-675,
WELLINGTON.

FIGURE 4-10
New Zealand's commercial broadcasters must abide by the nation's established standards for operation. Courtesy of the Advertising Standards Complaints Board.

tising income, in addition to user fees and government subsidies. The ABC has a further interesting source of income: auxiliary projects such as concerts given by ABC's six national symphony orchestras. Private stations are completely dependent on their advertising revenue.

New Zealand follows a similar pattern, with user fees and advertising as the main sources of income for Radio New Zealand and Television New Zealand, and advertising as the source of income for private stations. As noted earlier, in the late 1980s and early 1990s New Zealand was in the process of selling off many of its publicly owned enterprises to private owners, including foreign investors. The increase in private stations has resulted in increased advertising and concomitant income for capital for new stations (Figure 4-10).

Because New Zealand is considerably smaller than Australia, cable penetration is easier and pay-cable channels, mainly carrying satellite programming, are beginning to bring in even further income. Australia's vast size requires satellite transmission for network coverage, and in 1993 it added to its base of income by licensing two new satellite pay-TV channels. Both Australia and New Zealand have found production costs to be so high as to inhibit the amount of original programming needed to fill all the air hours, and both rely to a great extent on the importation of foreign programs. To pay for these programs, a barter system of advertising is sometimes used, and, in one New Zealand project devoted to international television programming, funding has been raised from foreign investors in exchange for a percentage of ownership in the project.

While the smaller countries, mostly individual islands or island chains, follow similar patterns of financing, their principal funding comes from user license fees, government appropriations, advertising, and foreign investments, or combinations of two or more of these sources. Their dependence on Australia and New Zealand offers them an additional source. Australia and New Zealand frequently provide surplus broadcasting equipment to places such as Fiji, Samoa, and the Cook Islands, thus enabling some of the smaller, poorer nations to have at least a minimal radio and/or television operations.

5 Broadcast Programming Worldwide

Depending on the kind of system on which it is aired, programming represents the bottom line of broadcasting. In systems relying on commercial advertising for their support, the value of programming is based on how many viewers any given presentation attracts, which in turn translates into rating points and income for the network or station. In systems operated by public corporations that rely on sources of funding other than advertising, the purpose of programming is oriented principally to the educational, social, and/or political purpose of the operating entity. No matter what the system, however, the content of programming can be extremely powerful, from providing chewing gum for the ears and eyes and anesthetizing a nation into couch potatoes largely oblivious to the crises of the times, to energizing a nation into dissatisfaction and political action by affecting the public's beliefs and feelings. Depending upon one's point of view, television and radio programming can be used for good or for evil.

In 1993 Shimon Peres, foreign minister of Israel, was quoted as saying that the greatest threat to Israel "isn't a military invasion, but a cultural invasion, and cable television is more dangerous to our identity than the Intifada." A 1993 international study comparing students' knowledge of math and science suggested that excessive television viewing might be an important factor in the poor showing of U.S. students. In the United States 22 percent of 13-year-olds watched more than five hours of TV a day, exceeded only by the youth in Scotland. While 79 percent of the young people in Italy and 64 percent in Spain reported doing more than two hours of homework each day, the figure was only 29 percent in the United States.

Conversely, for some students TV has added an extra dimension to learning, just as radio did some decades ago. Television and radio have given all people greater knowledge of the world about them, including new cultural concepts and information about other people and nations, establishing a base for understanding and peace as differentiated from ignorance and fear. At the same time, however, the media have given many people misinformation and false expectations through programming such as sitcoms and music videos. There are still people who insist that the quality of television is so low that they refuse to have a TV set in their homes. Contrarily, because the percentage of so-called cultural "junk" in print probably exceeds the percentage of "junk" on television, does one presume that those same people have burned their library cards to protect themselves from low-quality literature?

Whatever the merits of the media, the fact is that in the 1990s video and audio reception continue to expand rapidly throughout the world, with the continuing development of new and more efficient means of signal distribution. This growth means more channels and more hours to fill, and consequently, much more programming. Countries throughout the world have dealt with the volume and types of programming needed or wanted by their populations in different ways. One common factor is the increase in program exchange through satellites and an increase in co-productions among two or more nations.

NORTH AMERICA

United States

Programming in the United States is reflected in programming throughout most of the rest of the world. Popular U.S. shows, specifically sitcoms, night-time soap operas, and adventure/police dramas, are among the most popular programs in other countries. Internationally, some are picked up directly through DBS, others are distributed through cable systems, and some are dubbed for cable or terrestrial distribution.

Most programming in the United States is oriented to the commercial bottom line: the highest ratings and, therefore, the lowest common denominator. Programming on TV broadcast stations, for this reason, has virtually all been aimed at majority audiences. Only where individual stations may serve large minority or special-interest or needy populations has there been so-called niche programming on terrestrial stations. Not so for radio. When television took all of radio's prime-time entertainment programming in the early 1950s, radio's survival depended primarily on music formats and focused programming, with individual stations aiming at defined segments of their air population. A radio station that could garner and maintain even a small piece of the audience and advertising pie in a given community could not only survive, but prosper. Television, on the other hand, especially the networks, sought the widest, most general, largest audiences.

Cloning became a key. If one network was successful with a particular format, such as "Hill Street Blues," similar format "cop shows" followed on other networks. When "L.A. Law" became successful, a spate of lawyer dramas followed. Almost every sitcom can trace its roots to "I Love Lucy." The success of "Seinfeld" in the early 1990s resulted in a number of copycat adult sitcom series. The "Donahue" afternoon talk/interview program led to a host of highly successful copies. That is not to say that these program clones are lacking in quality; some are excellent examples of how good writing and production can make the most of TV's artistic potentials. However, most TV fare, original and clones, is considered by most critics to be at a low level, both intellectually and artistically. Nonetheless, the LCD programs appeal to a broad spectrum, and have become highly viewed in many other parts of the world, as well.

The U.S. broadcast system is primarily a commercial, privately owned one, with a relatively small but active public broadcasting component. The noncommercial television and radio channels are not dependent on advertisers for their support and were charged by the Public Broadcasting Act of 1967 to provide programming alternatives to the usual commercial fare. To a great extent public TV and radio stations have done so, with drama, music, dance, documentaries, and other formats, including news and entertainment, generally a cut or two above that found on commercial channels. Because public broadcasting does not have the funds of the commercial field, production values are sometimes of lower quality.

Even though noncommercial public stations do need to maintain an audience to merit continued support from their communities, and while their responsibility is to make alternative cultural and educational programs, they must also program in such a way as to attract as large an audience as they can. Therefore, public broadcasters, too, do not serve as many minority groups or needs as they might. Public broadcasting in the United States does not attract as many viewers as does commercial broadcasting, with average ratings running only about a 2—that is, only 2 percent of all U.S. homes with television—compared with five and six times that for commercial networks and stations.

The advent of cable provided an opportunity for additional program content and formats. To some extent cable networks function like radio stations, not able to compete on an individual basis with broadcasting networks, but seeking to obtain a sizeable share of a targeted audience. For example, the Black Entertainment Network (BET) on cable provides programming to and about African-Americans that is not available from any other source. The Discovery cable channel provides educational programs that attract viewers of all ages. The Nickelodeon cable channel carries entertainment programs, especially old sitcoms, aimed mainly at children. A system of special fee channels on cable provides some targeted entertainment, such as the Disney Channel and Home Box Office. In addition, there are pay-per-view channels principally devoted to sports events and movies. A significant phenomenon, anticipating interactive TV, has been the growth of shopping channels, whereby people at home literally watch displays and/or demonstrations of products that they can order via telephone.

As cable and VCR use has grown, broadcast network programming has lost audiences. Whereas in the early 1980s broadcast networks garnered some 90 percent of TV homes during prime time, by mid-1995 they had only 57 percent. This competition has led to greater attempts to attract broad-based audiences. New formats, such as "reality" shows, multiplied in the early 1990s, following the immediate success of a couple of reality programs, one violent, dealing with actual crime, and the other comedy, dealing with peoples' foibles. In the United States, format fancies periodically change. At different times the highest ranking shows have been westerns, then quiz shows, then comedy

and music variety shows. By the early 1990s, however, few of these formats were around in prime time, although the success of several western-oriented programs suggested a possible revival of the popularity of that format. In addition, cable took up the music slack with music videos, and audience participation quiz shows led non-network syndicated programs in the ratings.

Occasionally a format or program becomes so successful that it not only is exported internationally, but it affects the production of similar formats elsewhere. CNN is a case in point. A look at news programs all over the world in the mid-1990s shows that most of them have directly copied the CNN format used in its satellite-delivered, highly viewed world news program. In comparison to the previous news formats in many countries, the CNN worldwide format offers short, quick, no-depth startling headline news. This criticism is heard from journalists in many parts of the world. Yet, it is probably more of a commentary on other U.S. news programs than on CNN. CNN news generally provides more substance and a better selection of significant, as opposed to frivolous stories, than most other news programs in the U.S.

One growing area in the United States is Spanish-language programming. With an estimated 15 percent of the U.S. population of Hispanic descent, most of whom speak Spanish, and with several geographic areas, such as New York City, Southern California, the Southwest, and Florida with large Latino populations, there is a huge market. Many broadcasters have stated that they orient their radio programming or specific television shows toward a non-majority or special interest group because "we want to get a piece of the pie." Cable has made this more possible in television. In 1993, for example, one of Latin America's largest distributors of Spanish-language programming, Telemundo, began expanding its programming produced in the United States for American audiences. Comedy, variety, interview, talk, and sports shows, among others, were being distributed via cable and independent TV stations. Telemundo even had a rock music show in Spanish produced by MTV and a nightly news show in Spanish produced by CNN.

As technology grows, so do the expectations for the information superhighway, whereby homes throughout the United States will not only literally be able to receive hundreds of video signals, but will also be able to establish interactivity with many of them. This suggests the need for not only much more programming, but the development of new and different niche programming as well.

Radio is principally music on FM, mostly all-news and talk stations on AM, and some special interest stations, such as religious, community, and ethnic on both FM and AM. Music styles change, and from year to year one will find greater or lesser interest in stations carrying music formats such as country, top 40, middle-of-the-road, classical, easy listening, adult contemporary, and various types of rock. A format change developed in the late 1980s based on technology. With the truer sound for music on FM, AM stations began to lose listeners. In an attempt to find a niche, many AM stations took advantage of a sudden spurt of interest in talk-show hosts, especially those who could find an approach that titillated and/or outraged audiences. Many stations stayed in the black in the early 1990s by switching to talk. However, with the expectation of digital radio transmission before the end of the decade, AM and FM radio may (there are some skeptics) have an even playing field, which would result in significant changes in programming.

While many may regard it as tongue in cheek, a comparison of U.S. TV program quality with that of some other countries tends to validate the comment of TV critic John Crosby many years ago when he decided to end his column at the *New York Herald-Tribune.* Crosby said, when ceasing his coverage of commercial television, that it was not the commercials he minded, it was the programs that he couldn't stand any longer.

Canada

Canada fares somewhat better in program quality, but not nearly as well in commercial success. A key problem is the audience's interest in light entertainment, which results in the importation of many U.S. programs. Privately owned stations can maintain their ratings and advertising revenue much more cheaply by buying U.S. shows than by producing comparable shows of their own. In fact, Canada purchases more U.S. programming than does any other country. However, this has led to a problem with ethnic and national pride. Because of the large French-speaking population, many U.S. programs are dubbed into French, alienating many viewers who would prefer to see French-language original programs. As a consequence, and because the Quebec government has effectively used television for its regional nationalistic purposes, many French-language sitcoms are produced by stations in Quebec.

The Canadian Broadcasting Corporation (CBC) attempts to counter the influx of culturally deficient foreign programming by continuing the high quality of its own productions. While CBC offers news, documentaries, anthology drama, educational programs, and experimental programming, in order to maintain a competitive position and justify its existence economically it also carries entertainment such as police shows, sitcoms, adventure, family dramas, and mysteries. In 1992, to enhance its competitive status, CBC changed its evening schedule to include two hours of so-called adult programming from 10 P.M. to midnight each evening.

The private systems in Canada are comparable in programming to those in the United States, with national networks feeding local stations via satellite. Sports and entertainment dominate. While local affiliates in Canada take much of their schedules from their networks, including importations such as "M*A*S*H" and "Cheers," they do tend to do more with local news, documentaries and information programming than do local stations in the United States.

Even though Canada has restricted the number of U.S. broadcast stations that may be carried on Canadian cable systems, some 75 percent of cable programming is foreign, most from the United States. While Canadians near the U.S. border can pick up many U.S. stations, those further north cannot, and those who can afford to do so not only have cable, but DBS reception.

Cultural pride, however, loses to economic exigencies. As in the United States, Canada has university stations and public access cable operations, but the funding for these endeavors is insufficient to permit production of competitive quality programming. Even many radio stations import U.S. syndicated radio shows, such as Talk Net and Paul Harvey, although these importations comprise only a small part of any given daily schedule.

Central America

In Latin America programming is largely dependent on imports from the United States, Argentina, and Mexico, with additional programs from Spain and Portugal. In the early 1990s as much as 50 percent of the programming throughout Central and South America came from the United States. This number began to decrease in the mid-1990s as more countries produced more of their own shows. Satellite distribution through cable and DBS has expanded programming into the region from other countries throughout the world, especially from Spain. Argentine television distribution, for example, includes channels from Brazil, Chile, France, Italy, Korea, Mexico, Spain, the United States, and Venezuela. As the economy in any given country has allowed the development of more production facilities, dependence on foreign programming has decreased.

The most important home-grown production in Latin American countries is the "telenovela," or soap opera. Some countries have been able to develop novellas through co-production; for example, Venezuela's leading novella producer is associated with the U.S.-based Warner Brothers company. Mexico has developed telenovelas that have achieved the highest ratings, establishing a greater sense of rapport with Latin American viewers than the dubbed U.S. soaps or sitcoms. Most Latin American governments are careful to control or at least keep watch on content, and news and documentary imports from other countries are not as common as imports of entertainment programs. In addition, many Latin American countries have protectionist rules, but not for programming as much as for commercials. For example, both Mexico and Venezuela prohibit the importing of foreign-made commercials.

Multinational companies are increasing as distributors in the region as the number of viewers—and prospective customers for advertised products—increases. The U.S.'s American Broadcasting Company, for example, formed a group called Latin American Television International Network Organization (appropriately acronymed LATINO) over twenty-five years ago to sell its programs to Latin American television systems. Although such region-wide distribution systems still exist, in the 1990s the Latin American stations—as noted earlier, they are growing in numbers of private owners and in freedom from abusive government control—increasingly purchased programs on an independent basis, attending programming sales meetings in the United States and contracting individually for the purchase of series and individual shows.

Still, some networks have grown, rather than faded. Turner Broadcasting System introduced TNT to Latin America as the first satellite-cable-distributed, commercially supported, subscription service in the region. Turner offers programs in Spanish, Portuguese, and

English, including materials from WTBS, the Turner Atlanta broadcast station, and from CNN International. In 1992 Turner started a cartoon network which has made strong inroads in Mexico, other Central American nations, and the Caribbean. And in 1993 MTV inaugurated "MTV Latin America" to join its "MTV Asia," "MTV Australia," "MTV Brazil," "MTV Europe," and "MTV Japan" satellite distribution. MTV programming has been popular in Latin America for years through syndication. One of this book's authors remembers seeing one of the first music video programs in Latin America on a store-window TV set in Mexico City in the early 1980s. Crowds of young people packed the entire block in front of the store, turning the event into a mass rally for more music videos. Spanish-language movies also constitute important programming in Latin America, with many of the films coming from Spain.

As noted earlier, companies such as Telemundo produce programming for Spanish-speaking populations all over the world. A substantial amount of imported programming in Latin America comes from Spain. Mexico's media giant, Televisa, produces and exports internationally some 20,000 hours of TV programming each year, with an emphasis on Spanish-language soap operas.

No matter what the political system, whether democratic as in Costa Rica, totalitarian as in El Salvador, Honduras, Guatemala, Haiti, and others, or socialist as in Cuba, the programming is generally the same, although the specific content of a given program type may differ.

Mexico

In Mexico itself Televisa has gradually faced competition with the authorization of more private, individually owned stations. Nevertheless, programming competition is limited by the necessity for government approval not only of programs produced in Mexico, but of imported programs.

Guatemala

Guatemala's programming is representative of that in the region but is even under greater control, not only by the government, but by foreign companies such as United Fruit. That corporation at one time was not only instrumental in the development of radio in the country, but set up a network of stations in the region. Guatemala's television programming is mostly imported, but includes mandated educational and cultural programs.

Haiti

Haiti's programming was under strict control of the military after the coup against President Aristide; broadcasters were not only harassed but allegedly killed if they and their programming were considered threats to the new regime. All news and public affairs programming was oriented to the support of the government. Television is not yet very expansive in Haiti, with radio the principal communications medium. As in other countries, music constitutes the main audio programming. With the restoration of Aristide to power, the political orientation of the media changed, but programming remained under effective control of the government.

Cuba

Cuba's radio and television systems have no official censorship regulations, but since both are regulated and principally operated by the government, they support government policies. The two largest radio networks divide program types: Radio Reloj offers a news format, and Radio Rebalde has music, news, sports, drama, and cultural shows. Other radio networks include popular and classical music, and a station devoted to information for tourists. News dominates television programming, followed by films and children's programs, and includes public affairs programming, sports, music and variety shows, soap operas, and comedy. InterTV principally offers entertainment, and Canal Tele Rebele offers news, public affairs, and cultural programming. Economics limit Cuban-produced TV shows, and the most popular programs are the same sitcoms that get the largest audiences in other Latin American countries.

With the growing independence of stations in Central America, new indigenous formats have emerged, quite different from the foreign formats that were used as the basis for original productions. For example, a number of stations produce original dramas, their own musical events, and even circuses.

SOUTH AMERICA

Programming is similar among South American countries just as it is in Central America and in Latin America as a whole. Local programming is important to the scattered populations and the regional nationalism in a number of countries, although the central governments maintain strong control over network content and, in some cases, over that of local stations as well. Entertainment tastes perfer soap operas dealing with themes close to the viewers, and for comedy, particularly sitcoms. Much programming is imported, ranging from 30 to 70 percent, frequently depending on the economic viability of a given country to produce its own programming. As DBS and satellite-fed cable distribution grows, more programming of an international nature is reaching the people of the region.

Brazil

Although the only Portuguese-speaking country in South America, Brazil is highly cosmopolitan with a population representing all parts of the world and many diverse languages. Sao Paolo is one of the largest cities in the world with some 13 million inhabitants, many multilingual. Yet in its communication practices, including programming, the city is representative of all the other countries in the region. First, its programming is subject to government control, despite the fact that it has both public and private systems. Brazil is, in fact, one of the few nations in Latin America with elaborate and stringent mass media regulations. While radio is the primary media source for most Brazilians, as it is in other countries with a large number of poor people, in recent years television has begun significant penetration and impact. Also, as in other countries, while radio has become mostly music and talk, television has moved from primarily news and information toward more entertainment, with soap operas and comedy becoming the most popular fare. Further, as in other Latin American countries, commercial ads dominate the airwaves, accounting for 25 percent or more of air time. Because of the many different foreign groups in Brazil, there is probably more foreign language programming than in most other countries, and radio, for example, includes a substantial amount of music from other areas in the world. Yet, one of the frequently heard complaints about Brazil's broadcast systems is that not enough programming is oriented toward the many different ethnic groups in the nation, and that there is a tendency to ignore the country's cultural diversity. The Globo conglomerate dominates Brazil's broadcasting and does substantial production for export. Globo's soap operas and miniseries are the most popular programs in the country. Brazil imports about half of its video programming. Because of its size, Brazil was one of the first countries to experiment with satellite broadcasting, and it now uses satellites to cover the entire country with broadcast signals. In Sao Paolo especially, cable has grown and carries a number of foreign and international channels, including U.S. favorites such as HBO, CNN, ESPN, and Showtime. Also as in some other countries, the church plays a role in communications, and one of the major networks is principally religious in programming, with its news and public affairs programs not only highly rated in Brazil, but with its signal imported into neighboring countries.

Argentina

The second-largest country in South America, Argentina is almost a duplicate of Brazil, a further micro-example of communications in the rest of the region. As it becomes more cable-connected, Argentina's programming becomes more international in scope, including the U.S.'s HBO, CNN, and Cinemax, Korea Television, Mexico's "Eco" service, "Telven" from Venezuela, RAI from Italy, Spain's Television Espanola Internacional, and channels from France and Chile.

At the same time, locally produced soap operas and sitcoms are the most popular fare, and although almost half of its programming is imported, only one foreign program finished among the top-ten rated shows in 1993—the U.S.-produced animated show "The Simpsons." Radio, as in the rest of the region, is mainly music and news, with "Top-100" hits stations among the most popular.

Colombia

Even in a smaller country such as Colombia the programming pattern is the same. Television is used mainly for entertainment purposes, while radio is the medium for music, news, and information. A major difference between its radio programming and that of some other nations has historical roots. Colombia's early use of radio was to bring formal and informal education to its rural populations, which had little other opportunity for learning. Hundreds of thousands of transistor radios were given away by progressive political groups and by religious organizations and locked onto frequencies that broadcast such materials as housing repair, nutrition, health, history, and geography. Government stations today continue to provide such programming, while the private stations concentrate on music, sports, and news.

Television in Colombia is government operated, but with air time sold to commercial operators on two of its channels. Its programming reflects the entertainment formats of other Latin American countries. Many shows are imported from the United States. Nevertheless, over half of the production budget in TV is used to produce the highly popular novellas—the soap opera miniseries. However, with the government's tradition of use of the media for education, Colombian TV includes a fair amount of cultural, information, public affairs, and children's programming. In the early 1990s Colombia's TV systems began to increase production of such programs for regional and international distribution. With the complete privatization of television in the country anticipated in the mid-1990s, strengthened and expanded networks and foreign distribution are expected to increase dramatically.

Venezuela

Venezuela's television system is indicative of what may be a coming trend for smaller countries without the resources for local production. Some 90 percent of its TV programming is imported, principally from Argentina, Spain, Mexico and the United States. Its station managers, therefore, must work closely with the national distributors, who use a combination of satellite, cable, and local stations to reach viewers. With little local production to serve specific needs, the preponderance of programming is the most popular fare, such as soaps and sitcoms, to meet the perceived LCD.

Peru

Peru similarly reflects the region's programming in radio and television, mostly music and news for the former, and films, novellas, sitcoms, miniseries, and comedy for the latter. Also reflecting the unbridled commercialism on Latin

American stations, Peruvian radio devotes some 30 percent of its total time to ads, and TV some 40 percent. Although by U.S. standards imports are high, Peru's regulations require a minimum of 65 percent of programming and 100 percent of advertising on radio to be domestic. Despite the government's efforts to emphasize domestic TV programming, the regulations are not enforced because of the high costs of production. Therefore, in the early 1990s reports indicate that some 70 percent of TV shows were being imported, mainly from Argentina and Mexico in Latin America, and additionally from the United States, the United Kingdom, and Germany.

Chile

Chile is an example of what can happen when privatization threatens the existence of the government-run system. In an effort to boost its audiences and influence, in 1993 the government station National Television (NTV) began to go head-to-head with the private stations by running a competitive novella. It suddenly surged ahead in the ratings. Its success prompted some other nations' government stations whose traditional programming was being overwhelmed by the light entertainment of new private stations to consider the extent to which government stations—presumably still responsible for cultural, information, and education programs that are not usually provided by private stations—should adjust their programming to include popular fare in order to maintain their audiences.

EUROPE

The importing of foreign programs, especially from the United States, led to the imposition of quotas on program imports by the European Union in 1990. Many countries believed that both their cultures and economies were being hurt by such imports. They argued that U.S.-made LCD programs such as violence-prone dramas and inane slapstick sitcoms appeared to be harming the cultural and social fabric and artistic standards of their individual countries, and that their own TV production industries were being financially hurt by the competition. The quotas were affirmed by Europe in the 1993 General Agreement on Tariffs and Trade (GATT), in 1995 the EU closed loopholes that permitted a bypassing of the 50 percent rule and negatively impacted the local TV and film production industries. The original "Directive" was adopted by the European Commission's Council of Ministers in 1989 under the title "EC Council Directive Concerning the Pursuit of Television Broadcasting Activities." It specified that where practicable the majority of broadcast time should be reserved for "European works" with the exception of news, sports, advertising, and teletext. The Directive also regulates advertising time and ad placement, strongly recommending that commercials not be placed within programs. It deals as well with ad content, such as prohibiting the promotion of tobacco products and prescription medical products, and limiting alcohol ads. The Directive requires the restriction of programs that may harm the morals of minors, especially shows containing violence and pornography. The document guarantees the right of reply for any individual who is the subject of a personal attack in the media and requires that "broadcasts do not contain any incitement to hatred on grounds of race, sex, religion, and nationality." Needless to say, each member country has its own interpretation of the specifications in the Directive, and enforcement of the provisions on a standardized basis across Europe is virtually impossible.

At the beginning of 1994, U.S.-originated programming comprised 35 percent of the TV broadcast time in Germany, 28 percent in Spain, 25 percent in Italy, 22 percent in France, and 15 percent in the United Kingdom. The European countries have maintained that television is a "cultural commodity" and one French director, Jean-Jacques Beneix, hailed the GATT policy as "the birth of a . . . Europe which is acquiring a second dimension." On the other hand, U.S. Motion Picture Association president Jack Valenti stated that the protectionism has "nothing to do with culture," stating that European soap operas and game shows dominate the European airwaves anyway.

While the GATT agreement provides protection for European programming, the European Community (which became the European Union in 1994) agreement on cooperation threatened such protection within individual countries. The EC Directive of "TV without borders" removed restrictions on signals from one country to another. Some smaller countries, such as Belgium, find themselves deluged with stations from all over Europe that threaten to almost obliterate their own national systems. As noted earlier, Luxembourg's CLT and France's TF1 already began to dominate the viewing numbers in other countries. The stable or growing economies of most European countries in the 1980s accommodated the advancement of technology, and most nations became internationalized through satellite and cable growth. Nationalism necessarily became integrated with internationalism. While individual countries attempted to keep out programs of undue violence and other content incompatible with the given nation's cultural and educational goals, they also were growing more averse to censorship, as political entities in many parts of Europe became more democratized and decentralized. In his book *Satellite Television in Western Europe*, Richard Collins expresses concern with the degree to which satellite transmission will harm national cultures through carriage of programs deemed not only violent but in many cases pornographic. Concern has also been expressed over the uses of TV for political purposes, especially, as has happened in the United States, television's manipulation of the election process by favoring some candidates and

ignoring others. Carla Brooks Johnston, in her book *Election Coverage*, indicates the difficulties for fair press coverage of elections caused by changes in governments and differing concepts of democracy by different countries' media. She provides guidelines for broadcasters, with a Russian edition by the Moscow State University Press designed to help broadcasters new to democracy understand how to avoid subverting the new political process. A Council of Europe publication, *Aid for Cinematographic and Audio-visual Production in Europe*, reports on the growing cooperative work by writers and producers in many countries, attempting to find a quality international product that does not threaten individual national needs.

Co-production among different nations' systems is growing. By pooling resources it is expected that all-European products may appeal to larger audiences. For example, in 1993 eleven countries established a new channel, Euronews, to cover all of Europe as counterprogramming to the American-viewpoint CNN news, which has achieved great popularity in Europe and elsewhere. Another example is a soap opera series, "Riviera," a 260-episode show co-produced by France, Italy, and Germany. Soap operas are highly popular in Europe, with U.S. soaps especially successful in southern Europe. The highest-rated program in Spain for a long time was an imported Venezuelan soap, "La Dama de Rosa." More and more individual countries, such as Italy, Spain, and Greece, have taken advantage of the soap opera fad and have begun producing their own soaps with great success.

There are some generalizations that may be made about European countries' similarities in telecommunications programming. Interestingly, there are very few dissimilarities, except for special interests of individual countries.

As noted above, one distinctive feature of Europe is that in contrast to any other region of the world, with the exception of the United States and Japan, all countries have advanced systems of communication technology. Advertising, although growing in most countries with the increasing numbers of private stations and cable systems, is still generally restricted in total time, placement, and frequency, maintaining by and large the integrity of individual programs. Prime-time programs approximate those in other countries, with drama, sitcoms, audience participation shows, and films especially popular. Because there are still well-operated public and government systems, quality drama and documentaries are still available. In fact, compared to the viewing habits in the United States, European audiences assiduously watch good drama and documentary shows. Public stations continue to provide a substantial amount of news and educational programming not found on the private entertainment-oriented channels. Because of cross-border programming, almost any European viewer can obtain programs in different languages produced by and dealing with different cultures. European viewers have a great variety of choices of programs. U.S. television continues to exert significant influence on programming, in part through the importation of its programs, and in part through the cloning in various European countries of popular U.S. program types, such as the evening soap opera, the prime-time sitcom, and the audience participation show.

One interesting difference among countries is the preference for dubbing or subtitling for foreign programs. In Holland, for example, some 80 percent of the public prefers the latter, while in Germany and Switzerland about the same percentages prefer the former.

United Kingdom

Although Britain is part of the European Union, its geographical and historical position allows it to avoid the interdependence and cross-border interchange of the continent. At the same time, it is influenced by the same factors that affect countries on the continent, and through its BBC it has great impact on the structure, organization, and programming of broadcast systems all over the world, including Europe.

The United Kingdom, too, is concerned with program quality and its effect on cultural and social mores (Figure 5-1). Prime Minister John Major stated that he believed much of the violence on the streets of the United Kingdom was due to the violence seen on television, especially in programs imported from the United States. The BBC continues to maintain quality programming in the United Kingdom, despite the tough competition from the private systems. It is one of the few broadcast operations in the world that at one and the same time can provide entertainment to the mass audience and high-quality programming to a highbrow audience. The Independent Broadcasting Authority (IBA), with its Independent Television (ITV) programming to over a dozen local franchisees, Independent Radio programming to a network of local stations, and DBS services, provides entertainment programs comparable to those in other countries. The mixture of national and local programs maintains the concept of service to individual regions and communities in Britain to a greater extent than one finds in commercial networks in most other countries. In fact, one of ITV's channels, channel 4, is especially oriented to niche programming to special-needs or special-interest audiences. A relative newcomer, it has pushed the other channels to competitive quality programming, with its inclusion of programs on gay life-styles, third-world countries environmental problems, shows oriented to ethnic and racial problems, and programs on highly controversial issues, as well as the most popular U.S. sitcoms, talk shows, and even American football. As Channel 4 and Independent TV made strong inroads into BBC television audiences, BBC in the early 1990s began to experiment with more innovative programming, although, like other public

FIGURE 5-1
Statistics resulting
from BBC surveys.
Courtesy BBC.

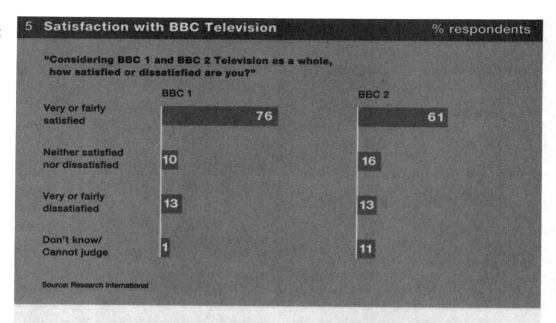

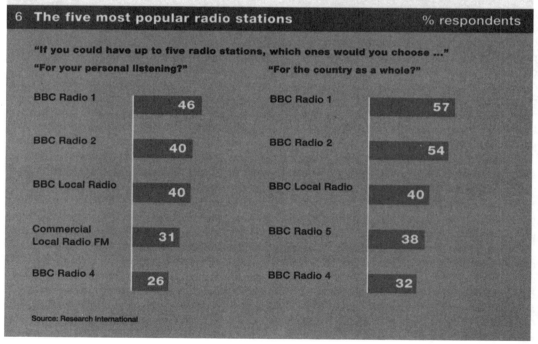

or government systems faced with the same kinds of competition, it has attempted to maintain the cultural orientation and high quality of its programs.

BBC Radio has long been the stalwart of audio programming, with its five radio networks in the early 1990s cover-

ing varying interests: Radio 1 has popular music, Radio 2 light music and general entertainment, Radio 3 classical music, drama, documentaries and discussion programs, Radio 4 concentrates on news, public affairs, and features, and Radio 5 is oriented primarily to chil-

FIGURE 5-1
Continued

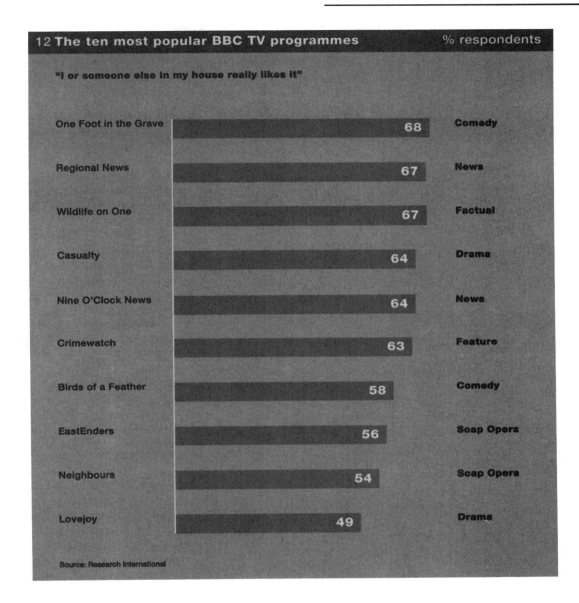

12 The ten most popular BBC TV programmes — % respondents

"I or someone else in my house really likes it"

Programme	%	Category
One Foot in the Grave	68	Comedy
Regional News	67	News
Wildlife on One	67	Factual
Casualty	64	Drama
Nine O'Clock News	64	News
Crimewatch	63	Feature
Birds of a Feather	58	Comedy
EastEnders	56	Soap Opera
Neighbours	54	Soap Opera
Lovejoy	49	Drama

Source: Research International

dren's and educational programs. The networks feed more than eighty stations in England, Scotland, Wales, the Channel Islands, and Northern Ireland.

Even the United Kingdom, however, has been invaded by the LCD talk radio shows that have become so popular in the United States in the 1990s. In 1995 the Talk Radio UK network was established, the first twenty-four-hour talk radio service available to commercial stations in the country. As reported by *Broadcasting & Cable* magazine, the network has "more lively and topical talk than [that] commonly broadcast by the BBC."

The BBC external radio service is known internationally and has long been considered the most reliable and thorough news operation heard in other countries. BBC television's external service provides satellite news feeds to most areas of the world, too. BBC TV also has a European service, which provides drama, soap operas, sitcoms, variety and comedy, science and natural history

Le programme Europe 2
Le plein de tonicité, d'émotion,
d'humour et d'impertinence !

"Depuis 1987, EUROPE 2 propose à
ses auditeurs le Meilleur de la
Musique dans un programme dont
l'exigence de qualité est reconnue par
tous. Nous sommes heureux de vous
présenter l'expression 1994 de cette
volonté."
Le Meilleur de la Musique...
Les meilleurs succès et les meilleurs
artistes d'hier et d'aujourd'hui.
Les souvenirs évocateurs de vos
meilleures années.
La Collection privée : chaque jour un
medley unique en son genre.

professionnels

COMITÉ DE DIRECTION

Gérant - Directeur Général :
Martin Brisac

Directeur Général Adjoint :
Yves Taieb

Directeur des Programmes et de la
Promotion :
Guy Banville

ADMINISTRATION

Secrétaire Général Adjoint :
Isabelle Andres

PROMOTION

Adjointe au Directeur des Programmes
chargée de la Promotion :
Delphine Marçais

Du lundi au vendredi

5 h 30   9 h 00
La matinale avec Vincent Neveu

5 h 30 : Titres de l'actualité
6 h 00 : Titres de l'actualité
6 h 30 : Titres de l'actualité
6 h 40 : Vive la crise de Zohra Taleb
6 h 54 : Les Guignols de l'info
7 h 00 : Journal
7 h 10 : Le Micro Caché de Pascal Sellem
7 h 24 : Les Guignols de l'info
7 h 30 : Titres de l'actualité
7 h 40 : L'Aventure avec Julien Pascal
8 h 00 : Journal
8 h 10 : Les Guignols de l'info
8 h 25 : Le Micro Caché de Pascal Sellem
8 h 30 : Titres de l'actualité + "Les Insolite
          la Presse" de L.Y. Giloux
8 h 40 : Les Guignols de l'info

Le meilleur d

En exclusivité !

En exclusivité :
Le Décompte Europe 2 : les titres
choisis par les auditeurs.
Le meilleur des concerts, des compila-
tions et des collections vidéos du
moment.
Les concerts acoustiques d'Europe 2.
Les concerts "Unplugged" avec MTV.
Les nouveaux talents français à
l'honneur : "Musiscope".
Toute l'actualité musicale en avant-
première avec "Backstage".
Le son magique des radios
américaines : "California".
L'essentiel de l'information :
22 éditions par jour.
Des rendez-vous cinéma, aventure,
consommation, etc.
La couverture des plus grands
événements sportifs et culturels.

Humour et personnalités :
Les Guignols de l'Info de Canal +
en exclusivité radio.
Richard Bohringer, Laurent Boyer,
Pascal Sellem et son "Micro Caché",
Marc Jolivet.

Une équipe de

Responsable des Relations Extérieures :
Corinne Marchand

PROGRAMMES

Adjoint au Directeur des Programmes :
Laurent Perrigault

PROGRAMMATION

Directeur de la Programmation
Musicale :
Christian Savigny

INFORMATION

Rédacteur en Chef :
Laurent-Yves Giloux

TECHNIQUE ET PRODUCTION

Directeur Technique France :
Michel Pierdait

Directeur de la Production :
Jérémie Blanc Shapira

REGIE COMMERCIALE/RRM

Directeur commercial :
Olivier Cantet

Europe 2
26 bis, rue François 1er - 75008 Paris
Tél. : (1) 47 23 10 63
Fax : (1) 47 23 10 59

9 h 00   12 h 30
La matinée avec Yann Arribard

9 h 00 : Flash
10 h 00 : Flash + "Le Monde change"
           de Michèle Fitoussi
11 h 00 : Flash
12 h 00 : Synthèse de l'actualité

12 h 30   14 h 00
programme local
14 h 00 : Flash

14 h 00   16 h 00
Lionel Safré
15 h 00 : Flash

16 h 00   19 h 30
Backstage avec Nicolas Du Roy
et Valli

16 h 00 : Flash
17 h 00 : Titres + MTV News de Valli
17 h 30 : Cinéma avec Patrick Fabre et
           Dominique Duthuit
18 h 00 : Journal
18 h 30 : La Collection Privée d'Europe 2
18 h 55 : Jeux
19 h 00 : Titres de l'actualité
19 h 29 : Titres de l'actualité

19 h 30   22 h 00
Programme local
22 h 00 : Synthèse de l'actualité

FIGURE 5-2
Weekly music
programming by
France's Europe 2.
Courtesy Europe 2.

programs, documentaries, and news
and public affairs programs to all of
Europe.

Western Europe

France

In the early 1990s France was one of the
fastest growing media markets on the
continent, with the fare on the public
and private television channels virtually
the same.

French radio devotes about 80 percent
of its air time to entertainment, about
15 percent to news and information,
and the rest to education and specialized shows. The several radio networks
and the more than 1,500 local stations

2h00   1h00
s chansons d'amour
vec Olivier Riou

00 : Les Unes de la Presse

h00   5h30
per-suite :
nuit toute en musique

Samedi

h00   11h00
ylvie Bariol

h 00 : Titres de l'actualité
h 00 : Titres de l'actualité
h 40 : Jeu "Encyclopédie Axis"
h 00 : Journal
h 00 : Flash et Europe 2 + de voile
de Catherine Chabaud

23h00   00h00
California

00h00   6h00
Super-suite :
La nuit toute en musique

Dimanche

6h00   11h00
Sylvie Bariol

7 h 00 : Titres de l'actualité
8 h 00 : Titres de l'actualité
9 h 00 : Titres de l'actualité
10 h 00 : Flash

11h00   13h00
Laurent Boyer et ses invités

11 h 00 : Flash
12 h 00 : Synthèse de l'actualité

A suivre...

**Europe 2 crée
l'événement tout au long
de l'année !**

Février :
Le Festival Débranché Europe 2
1er festival de concerts acoustiques
jamais produit en France.

Février-Mars :
Le Défi Europe 2.
Un jeu fou, fou, fou...
avec 100 000 F à gagner.

Mai :
Inauguration du tunnel
sous la Manche en direct.
Une production unique en duplex
Europe 2/BBC.

Juin :
Le cinquantième anniversaire
du Débarquement en acoustique.
Un méga-concert "Débranché" en
direct de Deauville.

Septembre :
L'école en musique.
Europe 2 fait chanter les écoles
de France.

Décembre :
Journée mondiale de lutte
contre le SIDA.
Europe 2 et ses auditeurs achètent
un appartement pour les malades.

Train de Noël Europe 2
Europe 2 sillonne la France en train
et distribue des jouets à tous
les enfants défavorisés.

e la musique

l1h00   13h00
e décompte Europe 2
vec Lionel Safré

1 h 00 : Flash
2 h 00 : Synthèse de l'actualité

13h00   19h00
rédéric Hubert

3 h 00 : Synthèse de l'actualité
4 h 00 : Flash
6 h 00 : Flash et Europe 2 + de vidéo avec
Patrick Fabre
6 h 00 : Flash
7 h 00 : Flash
8 h 00 : Synthèse de l'actualité
8 h 58 : Titres de l'actualité

19h00   20h00
Concert Europe 2

20h00   21h30
Programme local

21h30   23h00
Frédéric Ferrer

22 h 00 : Synthèse de l'actualité

13h00   18h00
Frédéric Hubert

13 h 00 : Synthèse de l'actualité
14 h 00 : Flash
15 h 00 : Flash
16 h 00 : Flash
17 h 00 : Flash

18h00   19h30
Frédéric Ferrer : Le tire-bouchon

18 h 00 : Journal
19 h 00 : Titres de l'actualité

19h30   22h00
Programme local

22h00   00h00
C'est beau une ville la nuit
avec Richard Bohringer

22 h 00 : Synthèse de l'actualité
00 h 00 : Synthèse de l'actualité

00h00   5h30
Super-suite :
La nuit toute en musique

FIGURE 5-2
Continued

concentrate mostly on pop music with
U.S.-type disc jockey formats, and on
news and interview shows. Call-in formats are popular (Figure 5-2). One of the
radio networks offers cultural programs
such as dramas, features, and shows
on the arts. Another stresses classical
music and news.

Television time is divided into about
65 percent entertainment, 20 percent
news and information, and the rest
principally to education and religious
programs and to advertising time. The
demand for entertainment by the public
resulted in a huge influx of U.S. LCD
(lowest-common-denominator)   shows,

and in 1990 France ordered a cutback on non-European programs to no more than 40 percent of TV schedules. Cable systems broadcasting in French are limited to 50 percent. However, French companies are increasing co-productions with European and noncontinental entities. Public stations do not use much imported programming, but private stations depend to a great extent on foreign shows.

TV formats during the day reflect a catering to demographics. Early mornings are spare, with mostly news and talk shows; afternoons offer long (one-hour) and short (quarter- and half-hour) soaps, dramas, talk shows, and some documentaries; late afternoon has children's shows, including cartoons, mini-documentaries, and rock videos; and evening prime time has drama, feature films, variety, sitcoms, evening soaps, and game shows. Nature programs are fairly popular. Like U.S. stations, French channels are highly competitive, with growing sophistication in counter-programming. Private stations show a lot of feature films, frequently highly rated and promoted foreign movies. The pay-TV channel, Canal Plus, has the most popular entertainment, including old U.S. sitcoms and prime-time soaps, talk shows, game shows, and a genre still popular in most countries, but generally absent from U.S. prime-time: variety programs.

Germany

With German unification, an adjustment had to be made by viewers and broadcasters to accommodate both the traditional viewing habits of east Germans and the fare produced for west Germans. The structure of regional stations governed by citizen boards promotes programming of a higher cultural quality than found in many countries with principally private systems. Nevertheless, popular fare is included on both TV and radio, with TV having such formats as feature films, sports, sitcoms, audience participation shows, and U.S.-type crime and adventure programs. While imported programs account for only a small percentage of German air time, some of the U.S. programs such as "Dallas" and Dynasty" have been among the most popular shows in the country.

Drama produced by the regional stations is usually of very high quality, sometimes adapted from novels, sometimes historical in nature, and usually with high ratings. Symphony concerts, operas, talk shows dealing with serious political, social, and economic issues, and high-quality documentaries, including many on nature and science, are also quite popular. The private channels, which in Germany are on cable, are principally devoted to light entertainment such as sitcoms and music videos and a U.S.-type approach to advertising. Private cable channels, in order to keep costs low because they are dependent solely on advertising, rely heavily on film packages, including a large number from the United States and Italy, plus packages of old U.S. sitcoms, music videos, and audience participation shows.

Both public and private channels carry documentaries, dramas, and variety shows. Both carry sports, although the private channels have put special emphasis on bringing in events that the public stations do not carry. Although cable is growing, the traditional regional public networks still command the largest audiences. One special area of programming cable has provided is that oriented to specific ethnic and nationality groups. German TV programming seems to follow the trends in the United States. In the early 1990s reality shows, especially those with a police theme, became among the highest rated programs in Germany. Drama consistently remains a favored format.

Commercials on the regional public stations are limited in time and placement so as not to interfere with program content. News is taken seriously, with national evening newscasts so closely followed that it is considered impolite to phone people at times when they may be viewing or listening to the newscasts. In addition, especially since reunification, German broadcasting has devoted large blocks of time to educational and cultural programs. While Germany imports some foreign programming, including that from satellites and cross-border reception, home-grown programs retain the largest audiences.

Public radio in Germany is frequently a cooperative affair, with regional stations exchanging programs or feeding programs into national distribution. Specified channels offer different formats: popular music, classical music, news and public affairs, foreign-language programming. Following the BBC approach, one radio channel is devoted principally to educational, informational, and cultural programming. Current affairs programs, particularly those dealing with the critical political and economic problems of the country in the early 1990s, garner large audiences. Private radio focuses on entertainment, principally mainstream popular music, with relatively little news or public affairs. A significant aspect of German radio that is surprising to many U.S. listeners, but represents the continuing practice in many countries of the world, is the retention, despite television, of formats such as drama, variety, and comedy. In fact, the regional broadcasting centers have elaborate radio studios equipped with various mechanical as well as electronic sound effects for the production of dramatic shows.

As indicated earlier, programming in the smaller countries of Europe is eclectic, reflecting imports from cross-border signals and international satellite transmissions. This has placed a heavy burden on public or state systems that emphasize cultural and informational programs, with the popular entertainment of foreign programming getting the principal attention of the audiences.

Belgium

Belgium, for example, gets most of its programming from the SKY channel, France, Italy, Germany, Holland, and Luxembourg. The bilingual nature of the country—French in the south, Flemish in the majority population north—makes it more difficult for the government to provide programming that serves and strengthens national needs. One method of raising funds to do so is through a special levy on cable systems, which carry mostly foreign programs. This tends to somewhat reduce the imports and at the same time provide production funds for endemic programming. Belgian advertising practices have followed the U.S. approach, affecting the quality and content of programming. As more and more outlets enter the market and compete with other stations for audiences, producers are inserting commercials into the most suspenseful and critical sequences to be certain that viewers will want to stay tuned. The formats that are most popular are similar to those in the United States, such as "The Tonight Show" type of variety, talk shows, sitcoms, action and adventure (the U.S. program "MacGyver" has been among Belgium's top-rated programs), soaps ("Dallas" has also been among the most-viewed), and game shows, such as "Wheel of Fortune."

The Netherlands

Although a small country, the Netherlands' unique system of allotting programming time results in the public channels carrying programs made in Holland and reflecting the constituent needs of the Dutch organizations producing the programs. The programming therefore encompasses many different types and formats, unlike the block programming inimical to the United States and many other countries.

As noted earlier, the Netherlands Broadcasting Foundation, Nederlands Omroep Stichting (NOS), is the coordinating body of the production groups and maintains a schedule of news, discussion, and children's educational programs. Informational programs account for a large proportion of air time. That is not to say that entertainment programs are neglected. While dramas have large audiences, sitcoms, variety shows, comedy, feature films, and audience participation shows are also popular. Many Dutch quiz shows are similar to those in the United States and are just as popular, consistently among the highest rated programs.

Even so, about half the total air time consists of foreign programs delivered via cable from satellite and over the border, with German and French channels easily receivable. It was mentioned earlier in this book that a channel received over the air from Luxembourg, RTL4, includes entertainment programming produced in Hilversum, the broadcasting center of Holland, including an interview-variety show similar in format and audience support to "The Tonight Show." The show usually is called by the name of its national personality hostess, Tienike, just as "The Tonight Show" was usually called the Johnny Carson show.

Radio in Holland is programmed in essentially the same way as television, by organizations representing citizen groups. For many years the government radio system concentrated on news and high-quality entertainment, giving rise to "pirate" stations off the Dutch coast that played rock and roll and other popular music, eventually forcing the official radio to include entertainment in its formats. Because of pillarization, Dutch radio includes many serious formats such as discussion, educational, informational, arts, and cultural programs.

The Scandinavian countries exemplify the strongest efforts to maintain quality cultural and educational programming in the face of increasing privatization of stations and satellite and cable distribution of LCD entertainment programming. In general, imported favorites are dramas from the United Kingdom and game shows from the United States. As in other countries, cable systems are dependent on advertising and therefore concentrate on the most popular entertainment programs.

Finland

The Finnish Broadcasting Company, YLE, began an effort in 1990 to retrain its writers and producers to make their traditional programming more entertaining without giving up program quality and content that strengthen Finnish culture. YLE's major radio formats are news and information, cultural and artistic materials, including dramas and features, educational programs, and light entertainment. TV fare is similar. Private and satellite-distributed programs are more reflective of popular entertainment. An interesting anomaly concerning feature films: the satellite channels carry foreign films, including well-known U.S. movies, that have been edited for general TV audiences; YLE, however, does not censor films, and even satellite aficionados turn to YLE for the same films they might see on cable because the YLE presentations are in their original entirety, without sexual and other scenes edited or deleted.

Sweden

Swedish broadcasting is somewhat similar to that of YLE, with local stations attempting to meet the special needs of their audiences, with an emphasis on formal and informal educational materials for labor and industry as well as for traditional young and adult learners. However, the government is greatly concerned with program materials that might be deemed harmful to the public. Contrary to the practice in most other countries, these concerns rarely include controversial materials on news, public affairs, and discussion programs. The concern is more related to violence, harmful advertising, materials promoting ethnic or racial discrimination, and libel. Also, contrary to the practices in most other countries, sex and nudity are not considered censorable.

Sveriges Riksradio does its programming in the same manner as the BBC. One network offers news and popular music, another provides classical music, and a third light entertainment, especially for distribution by local stations (Figures 5-3 and 5-4). All networks provide news and educational programming. Another radio network is devoted to educational programming, on both the formal classroom and adult learning levels. While the public TV system programs materials designed to strengthen the culture and education of the country, it includes a mix of selected entertainment shows. The private satellite channels concentrate on popular fare, including feature films, sitcoms, soaps, game shows, police and adventure dramas, and other favorite U.S. and U.K. programs. For example, some of the most popular programs in Sweden in the early 1990s were the U.S.-produced "Baywatch," "Jake and the Fatman," "Dallas," "Star Trek," "Happy Days," "The Untouchables," and "Laverne and Shir-

ley." As in the other highly educated Scandinavian countries, news programs are considered extremely important and listened to and watched assiduously.

Greece, Italy, and Spain are examples of countries that have a dual system of program distribution and program content. As noted earlier, soap operas are particular favorites in these countries, with afternoon soaps drawing higher audiences in some areas than prime-time programming, including feature films and variety shows. The afternoon "siesta" break may contribute to this. Local cloning of shows from the United States that draw audiences away from national productions has been a successful method of competing. For example, in 1993 the same kinds of reality shows that had become the new hit format in the United States were being produced

FIGURE 5-3
Swedish Broadcasting Corporation's programming data. Courtesy SBC.

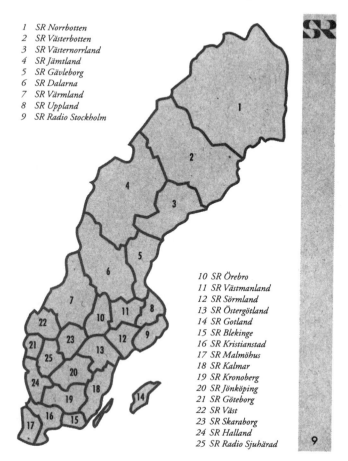

1 SR Norrbotten
2 SR Västerbotten
3 SR Västernorrland
4 SR Jämtland
5 SR Gävleborg
6 SR Dalarna
7 SR Värmland
8 SR Uppland
9 SR Radio Stockholm

10 SR Örebro
11 SR Västmanland
12 SR Sörmland
13 SR Östergötland
14 SR Gotland
15 SR Blekinge
16 SR Kristianstad
17 SR Malmöhus
18 SR Kalmar
19 SR Kronoberg
20 SR Jönköping
21 SR Göteborg
22 SR Väst
23 SR Skaraborg
24 SR Halland
25 SR Radio Sjuhärad

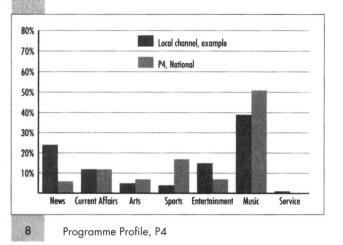

P4 – Service for a Mature Audience

P4 is made up of 25 regional, channels with local radio headquarters in around 60 locations throughout Sweden.

The 25 local channels have a particular responsibility for promoting local culture and music. Broadcasts are between 6.00 a.m. and 7.00 p.m. on weekdays, as well as on Sunday mornings. The stress has been placed on news and information via an integrated system of local, national and international news and current affairs broadcasts. During the evenings and weekends, the channels that make up P4 transmit network broadcasts nationally, including programmes from the Radio Sport service.

The aim is to achieve a daily listening audience consisting of at least half of the channel's target group, the over 35s.

Programme Profile, P4

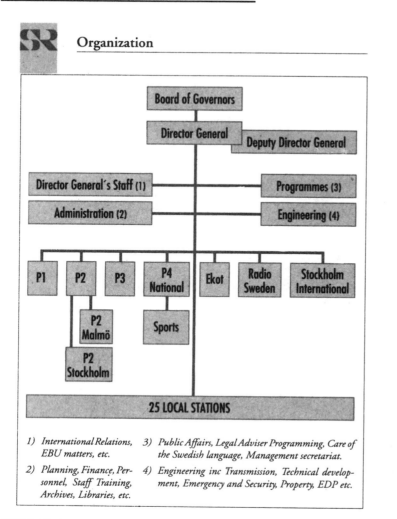

Organization

1) International Relations, EBU matters, etc.

2) Planning, Finance, Personnel, Staff Training, Archives, Libraries, etc.

3) Public Affairs, Legal Adviser Programming, Care of the Swedish language, Management secretariat.

4) Engineering inc Transmission, Technical development, Emergency and Security, Property, EDP etc.

FIGURE 5-4
Swedish Broadcasting Corporation's local station organizational chart. Courtesy SBC.

in Italy and Spain and receiving top ratings. In 1993 not one but two reality shows in Spain were patterned after the U.S.-made "Unsolved Mysteries," one with the same title and the other entitled "Code One." At the top of Spain's rating charts for a long time was a show entitled "Quien Sabe Donde"—"Who Knows Where?"— a program with a missing-persons format.

Spain

An interesting anomaly about Spanish broadcasting is that although the private stations are oriented more toward LCD entertainment than are the public stations, which remain under strong influence of the government, the private stations also provide more controversial news and public affairs programs than do the public stations. Both formats draw high ratings. Although the Franco dictatorship died with his death in 1975, the fascist tradition of censorship made the public stations wary, and they are still loathe to antagonize even unintentionally the government in power. In the entertainment area public stations have attempted to maintain high quality drama and cultural programs, but have been forced to compete with the private stations by counterprogramming and running commercially attractive shows. Hollywood feature films are among the highest rated presentations on Spanish TV. Game shows, soap operas, sitcoms, miniseries, variety shows, and sports dominate the airwaves, as in other European countries. The private networks have been able to offer the most popular programs, aided by Spain's excellent dubbing industry. Popular U.S. films and high-profile performer variety shows get very high ratings. Sports are extremely popular in Spain, and in 1989 the regional stations established a cooperative, the Federation of Autonomous Radio and Television Organizations (FORTA), to buy key programming, including an exclusive contract to carry the Spanish League soccer games.

U.S. sitcoms and action series are also popular, with "L.A. Law," "Cheers," "Murphy Brown," "The Golden Girls," "Moonlighting," and "Beverly Hills 90210" among the most watched. American movies are staples on Spanish TV. Some stations carry late-night porno films.

Yet, at the same time, all broadcasters are required by the government to carry programming that enhances Spanish culture and education, and a ratings board classifies programs into age group suitability, similar to the film ratings in the United States. Despite the importation of almost half of its TV programming, formats that lend themselves to local needs, such as audience participation and game shows, are produced by the individual Spanish networks.

Italy

Italy faced the same problems with increasing privatization in the 1980s, particularly with the emergence of media baron Silvio Berlusconi, who quickly dominated the media in Italy, much as Rupert Murdoch has done in other parts of the world. His stations—the three largest private TV nets in Italy—have benefited from his acquisition of rights to popular U.S. series, such as "Dynasty," and "Dallas." The importation and popularity of foreign programs led to government restrictions on the amount of foreign programs that could be broadcast, and the designation of minimum-percentage requirements for locally produced shows. When Berlusconi's private stations began to overwhelm the government network RAI, competition set in on two levels. On one level, both the government and many Italians were concerned that the popularity of foreign programs was vitiating Italian culture and identity. Government regulations have resulted in a serious cutback in U.S. programs, with films remaining the principal import.

The second level of competition was the production of programs by RAI and by the private stations, especially Berlusconi's, that tried to attract the most viewers. Since Berlusconi carried U.S. game shows, RAI developed its own versions. When Berlusconi drew audiences with giveaways, RAI invested more monies in prizes than did Berlusconi. While RAI eventually won back a substantial viewing audience, the costs of doing so resulted in subsequent cutbacks in its budget. When Berlusconi became prime minister in 1994 on a platform supported by fascist allies, some critics wondered whether he would use his position to quash any further competition to his own stations.

By 1991 Berlusconi's Finvest group was producing more TV fare than any other European organization. Cutting through the bureaucratic red tape that has hampered the efforts of many public stations and also many private stations under strong government regulation, Berlusconi was able to develop co-production agreements with other European countries in an effort to produce programs that could be widely distributed in many countries at once. As co-production grows, these broad-based entertainment programs will further eat into the audiences of public stations' cultural and educational offerings throughout the continent. Murdoch's efforts to establish a global system are matched by Berlusconi's plans to do the same thing. Will we find an ever-decreasing diversity of programming and points of view in countries, in regions, on continents, and ultimately throughout the world as one or two individuals gain ever-widening media monopolies?

Reality programming has been so successful in Italy that political pressure was brought to bear to cancel one program that dealt with real-life political events and invited callers to phone in with their opinions and ire. A leading program in Italy in the early 1990s was "Paperissima" ("Goof-Up" in English), a copy of the U.S. show "America's Funniest Home Videos." Drama series are oriented more and more to current affairs themes, with the highly popular "The Octopus" based on organized crime in the country. Even TV movies are becoming more oriented to reality themes, including political and social issues. Some producers and critics expect that this trend in Italy will move drama throughout Europe toward its own special format and style, different from and with increasingly less reliance on U.S. drama exports. Perhaps the growth of national and European innovative programming accounts in part for the unexpected failure of U.S. sitcoms in Germany in 1993. German-adapted formats of "Who's the Boss?" and "Married—With Children" did poorly, despite the fact that the U.S.-dubbed versions have been doing reasonably well in Germany for some time. Judging from the successes of British programs when adapted in the United States into "Sanford and Son" and "All in the Family," one might well expect the local orientation of a successful format to work.

Portugal

Portugal is an example of gradual changes in programming and structure when commercial stations begin to compete with government stations. In 1992, after forty years of state-operated TV, Portugal's first commercial TV station went on the air. As an indication that the transition would not be too great, the station's first director was the former prime minister of Portugal. The station, Sociedade Independente de Communicacao (SIC), began its programming with an emphasis on news, current affairs, and reality shows. Sitcoms, films, sports, and other forms of entertainment were secondary. The second independent TV channel went on the air in early 1993. As an indication that here, too, a change to instant LCD entertainment would not occur, the Catholic Church was authorized to be the principal operator.

Austria

Austria is one of the few countries that, as of the early 1990s, had not yet authorized private broadcasting. The government-operated system, ORF, continued to emphasize Austrian culture while at the same time seeking to hold on to an audience that could get foreign programs through satellite, cable, and cross-border signals. It maintained entertainment shows for about one-third of its schedule, with another one-third divided between information and educational and cultural programming. It stresses Austrian culture, especially in the programs of its regional stations, including local news, features, and regional folk music. National TV broadcasts devote about 30 percent of the schedule to news, information, and sports, about 25 percent to entertainment, including drama, about 20 percent to cultural programs, and about 10 percent to education.

Austrian Radio devotes about 40 percent of its time to entertainment, including popular music, about 20 percent each to news and information and to cultural and educational programs in the arts and sciences, and about 5 percent to target audiences such as nationality, ethnic and religious groups. About 10 percent of air time is taken up with commercials.

Greece

Greece is an example of a country where the introduction of commercial TV, even with its LCD programming, presents to the public new ideas and attitudes that one ordinarily expects to be stimulated by public TV. When private stations were authorized in 1989 Greece's two government channels offered high-quality drama, cultural, and educational materials. However, news and information was strongly controlled and controversial issues were avoided. Many critics believe that the influx of foreign programming and the cloning of foreign formats have exposed the public in Greece to materials it had not gotten from the government channels. The different lifestyles on TV entertainment programs have also opened up viewers to behavior patterns and the discussion of subjects that had not been dealt with openly before, such as homosexuality, domestic abuse, and AIDS.

Although prior to commercial TV the government channels did run some U.S. series, such as "Dynasty," its imports were principally along the lines of the Jacques Cousteau documentaries, and it emphasized—and still does—documentaries, educational and cultural shows, and government reports. In addition to the eye-opening entertainment programs, commercial TV introduced talk shows, and for the first time the Greek public could see interviews and discussions on the same subjects that titillated other countries, such as the Bobbitt penis and Harding-Kerrigan ice skating scandals of late 1993 and early 1994. Previously, all discussions of any such events had to be government approved. An example of privatization's impact was the series on the Greek Civil War that followed World War II, giving many Greeks for the first time information on what really happened and creating new understanding of the conflict between fascist and socialist ideologies in their country. Even the electoral system was affected, with commercial stations giving the voters more information on the various candidates and parties than they had ever gotten from the government stations, which promoted their own agenda. In fact, some credit this change with making it possible for the democratic-socialist government of Andreas Papandreou to displace the ruling conservative party. This is not to say that commercial TV was all or primarily a breath of fresh air for social and political information. In early 1994 the highest rated show on Greek TV was "Wheel of Fortune."

The two government TV channels have adapted to commercial competition, with one continuing to stress news, cultural and informational programs, and international films, and the other offering entertainment programs such as game shows, sitcoms, soap operas, and Greek films.

Whether the Greek experience can be replicated in other countries depends on the degree of government control of the public system and the degree of exposure otherwise to foreign programming through cross-border or satellite importation. In most democratic countries the introduction of private ownership has led principally to a greater scarcity of ideas and information because of a total concentration on money-making LCD entertainment. In totalitarian countries the commercial stations support government policies in order to continue their mutual existence.

Some think that the Greek experience can serve as a model for eastern Europe as its media become increasingly independent of government control for the first time in their histories. However, traditions die hard, and even where new systems and infrastructures have been developed, the outlook in 1994 was not as promising as many observers expected. For example, in Russia President Yeltsin's dictatorial seizure of control of the media when he was faced with political opposition; and in Poland, while broadcasting is largely free of government control, its subservience to the church in its programming practices. As new political systems, some of them striving for democratic bases, are in power struggles to become established, their leaders are aware of where power lies, as stated by Rusina Volkova, secretary to the Russian embassy in the United States: "There are two main powers in Russia: the army and television." It is understandable, although not necessarily justifiable, to see strict control of programming by the parties in power in eastern Europe in the early 1990s.

Smaller countries try to protect their cultures by limiting the importation of programs, but at the same time are limited by their size and economies from doing as much original production as they would like. For example, Switzerland, as affluent as it is, has been unable to meet all of its public's needs through the Swiss Broadcasting Company. Therefore, private stations, satellite, and cable have provided popular entertainment programming. Part of the Swiss problem is the great cultural and language diversity throughout the country, necessitating broadcasting in German, French, Italian, Swiss dialects, and English. In addition, signals from surrounding countries cover most of Switzerland.

Eastern Europe

Russia

Programming in Russia and the Commonwealth of Independent States (CIS) changes like the political weather. For example, at one point in the early 1990s the head of state television arbitrarily coped with the increasing presentation of controversial subjects on television by designating one channel as all-entertainment and another as the "president's channel" for the dissemination of news and information deemed suitable for the public. Discussion programs presenting various political viewpoints that were not openly discussed before became audience favorites. At the same time, entertainment programs that once had rarely been seen or heard, such as sitcoms, soap operas, and pop music, also became favorites. Even the previously staid and culture-information-oriented government TV channel, Ostankino, arranged to carry popular U.S. TV shows. The highest rated program in early 1992 was the U.S.'s "Disney Presents." Other favorite shows that year were "Murphy Brown" and "Perfect Strangers," and even before that "Dallas" and "Santa Barbara" garnered large audiences in Russia. Russian TV began to produce its own programs based on successful and popular U.S. formats. The most popular program in Russia in 1994 was a Russian adaptation of the U.S. game show "Wheel of Fortune." The second-most popular program that same year was "Simply Maria," a soap opera imported from Mexico. When the star of the program toured Russia, she received a heroine's welcome and was mobbed by admiring fans everywhere. In Russia and in other eastern European countries where an attempt at instant capitalism has resulted in instant poverty and gangster violence for many, escapist programs help the public get their minds off their newly shattered economies and disintegrating living conditions. Some observers suggest that the combination of bright colors and lovely homes in the programs and the traditional Russian tendency toward emotionality make soap operas natural winners.

Of the six Russian TV channels, only one in 1994 was privately operated, and the programming on all six was somewhat similar. The growth of foreign importation is evident in the increasing number of satellite dishes; those who can afford such reception can get CNN, MTV, and other satellite-delivered networks. When importation of satellite TV signals was authorized in 1990, a number of foreign programs, including news and entertainment, were made available to Russian viewers.

Russian state radio, Gostelradio, operates several networks, dividing programming along BBC lines. One network concentrates on news and information, another on music with occasional news reports, and a third on educational and cultural programs. The new independent stations try to offer alternative entertainment programming, but with the continuing political crises in Russia in the early 1990s many of these stations found that their principal roles were as alternative political stations, providing news, information, and discussions not permitted by President Yeltsin on the state stations.

Russian television still does make a special effort to emphasize culture, and the dance, classical music, opera, and similarly artistic presentations are among the highest quality of any in the world.

The other CIS nations operate their own systems and orient their programming along similar lines, under strict government supervision. Some of their programming is received via satellite from the major channels originating in Moscow; but reception of these channels depends on the relationship at any given moment between the receiving country and Russia. All, however, have publics eager for western fare, the light entertainment that they had rarely seen or heard before. Much of this light entertainment is imported from the United States, the United Kingdom, Germany, France, Mexico, and Brazil, in part because the economies of eastern European countries make it difficult for them to produce their own LCD programs. The degree of flexibility in programming varies from country to country, depending on each one's political configuration at any given time. While Turkmenistan, for example, maintains strict government control, and all news on the media is government news, Uzbekistan imports a variety of western shows, including cultural programs, MTV-type music shows, and U.S. sitcoms and drama series.

Radio is more independent, with stations such as Moscow Echo continuing to offer alternative programming, both in areas of entertainment and in news and public affairs. As in other European countries, radio in eastern Europe has not abandoned its potential and surrendered entirely to popular music, but continues to offer drama, variety, educational, and similar programming. But it should be remembered that in many eastern European countries popular music was the alternative programming, banned by many governments for many years. Therefore, pop music on eastern European radio, unlike that in the United States and many other countries, is considered fresh, alternative, independent programming.

Former Soviet Bloc Countries

In early 1991 a number of former Soviet republics and eastern bloc countries became members of the European Broadcasting Union and began receiving some of their news and public affairs programming from the EBU. This has spurred some countries to seek more depth, breadth and objectivity in some of their news reporting. Czechoslovakia (now the Czech Republic and Slovakia), Hungary, and Poland in particular opened their media to an exchange with western Europe. The networks in the Czech Republic and in Slovakia (although the latter's shaky political and economic bases have resulted in tighter government control of the media) carry entertainment, news, documentaries, drama, and films, with audience tastes leaning toward light entertainment. Some 40 percent of their programming is imported, and the stations attempt to meet the ethnic and nationality needs of the citizens by having programs in Czech, Slovak, Hungarian, and Ukrainian. In addition to the church orientation noted above, Poland does carry

over the educational orientation of its former regime with excellent drama, children's, and educational programs. In an effort to expand the contents of its system, Poland has encouraged submissions by independent producers. Programming in Hungary is somewhat the same, mixing new entertainment shows with news, political discussions, cultural presentations, and educational materials. Programs are produced in somewhat the same way they are in Holland, by government-recognized organizations and citizen groups. While in 1993 there was still only one private TV station in Hungary, its owner, Tamas Gyarfas, was recognized as the country's first media mogul and increased privatization was expected. Gyarfas's network, NAP-TV, began producing programs innovative in Hungary that attracted large audiences, programs that included a regular panel discussion of controversial political issues, and a late evening personality and variety program similar to the U.S. offerings, "The Tonight Show" and "Late Show with David Letterman." Hungary also is illustrative of trans-Europe entrepreneurship, in 1993 inaugurating a new satellite channel, Mozaik TV, which began sending arts and cultural programs, with an emphasis on feature films, to audiences in Bulgaria, the Czech Republic, Romania, and Slovakia, as well as in Hungary. As in other countries with multiethnic populations, Hungarian stations broadcast in Hungarian, German, Rumanian, and Serbo-Croatian.

Bulgaria, which like the other eastern Europe countries imports some 40 percent of its TV programming, offers similar programs but in 1990 added an interesting block of programs to one of its channels: programs in English, French, and German aimed at tourists. While Bulgaria retains its tradition of using the media for information and education, it imports many LCD programs, including U.S. daytime and evening soaps. As privatization of the media grew in the 1990s, so did advertising, with many commercials for goods imported from nearby countries. Romania also combines informational and cultural programming with new, west European-type light entertainment, in German and Hungarian as well as Rumanian. One of its most popular genres in the early 1990s was western movies and series from the United States. About 20 percent of its total programming, a higher percentage than in most western countries, is devoted to news and public affairs. In 1991, with the new blush of political freedoms, Romania provided a channel called Free Rumanian Television as media access for citizens to present their views. Albania took more time than other eastern European countries to move toward entertainment, and in the early 1990s the government-operated system continued TV fare principally oriented to information, documentaries, cultural, and science programs. Its importation of some 40 percent of its programming, however, increased the audience's appetites for light entertainment.

Lithuania is an example not only of the Baltic states but of other former Soviet bloc countries whose programming is still largely dependent on Russia's, and where the Russian language continues to serve a substantial part of these countries' populations. Radio broadcasts are also in Polish, acknowledging Lithuania's incorporation in that country at one time. When the political ties to Russia were dissolved, private TV stations began to appear, including satellite importation of foreign movies, sitcoms, soaps, and MTV. At the same time Lithuania continued to carry TV relays from Russia throughout the country. As the 1990s advanced, Lithuania and other Baltic states attempted to break off from communications dependence on Russia and to strengthen their own programming.

A number of countries outside of the CIS have acquired permanent residents from Russia and the Ukraine during the past several decades, and although the Russian language has been frowned upon and even discontin-ued in many schools as a result of political backlash, programs from Russian TV and radio are carried for these expatriates.

AFRICA

One of the concerns of third-world countries in Africa and in Asia is the misconception about their situations by most other people in the world, particularly those in western and Asian industrialized countries upon whom they depend for assistance in emerging from their economic strictures. Responsibility for these misconceptions is placed on news and public affairs programs that are rife with misinformation. A 1993 public opinion study in the United Kingdom by World Vision concluded that the "public has a grossly distorted view of the Third World." The media are in great part to blame because they concentrate on the dramatic events that are not necessarily representative of a given third-world country, and ignore the typical events that would provide a truer picture of a given country or region. For this reason many third-world countries are concerned about programming to a greater degree than are non-third-world countries. The latter may also present distortions about themselves through exported programming (i.e., many countries consider the United States a nation whose every city and town is fraught with violence and whose every citizen owns a palatial home and lavish material goods) but are affected only in reputation and not necessarily in their economic growth, as can happen to third-world countries. The other side of the coin is that in virtually all African countries there is strict government control and censorship, even where there are non-government public stations or privately owned stations. The news emanating from these sources is no more and probably less objective than the selected news impressions that many of these countries complain about.

As discussed elsewhere in this book, programming is oriented toward the wishes of the party in power and, in the Gulf States in particular, in accordance with the state Islamic religion and the interests and even peccadillos of the ruling royal family. While some countries, such as South Africa, permit light entertainment programming, any information presented over the media represents the viewpoints—indeed, is the propaganda—of the government in power. In countries where the media are totally state controlled and operated, programming is determined by political overseers or directors; in countries where programming may be initiated by the individual station or network, it is either cleared by the government or is self-censored under pain of imprisonment or, as has happened in many instances, execution. One must keep in mind, however, that the rationale for control and censorship is both political and economic, as described earlier, and that the government justifies such control by stressing the nation's need for common cultural values and social and political behavior that, in the government's eyes, strengthens the nation and assists its growth. The bottom line, in Africa and in every country in the world, including democracies such as the United States, is the need to manipulate media programming in order to keep the current government in power. As even the governments with the strongest controls put it, they are merely offering the media "guidance." Because of this political purpose, programming in African countries tends to be somewhat more oriented to news, information, and education than that in other regions.

Indeed, the development of radio and, later, television, was in many instances the result of specific and obvious political or social needs, even after independence was achieved from the colonial countries that initially imposed their communication systems on the countries they governed. For example, Nigerian television was established, according to some sources, because a western political leader was denied access to government radio channels and he therefore made an arrangement with a British company to found Western Nigeria Television. In Saudi Arabia the government was concerned that the inevitable introduction of television would provide the public with information and images it did not deem appropriate, so it started television under its own control and orientation. Radio, initially developed for colonial power propaganda and colonial officials' entertainment, remained an important entertainment medium for the people of the newly independent nations, who, in many cases because of religious taboos, poverty, and the designation and treatment of women as chattel, had no other public outlet for entertainment. Some countries, such as Algeria and Morocco, found that radio provided needed links with neighboring countries sympathetic to their struggles for independence or post-independence stability. The emperor of Ethiopia, Haile Selassie, copied the lead of his invader—Mussolini and the Italians—in the 1930s by using radio to rally support for his retention of the throne.

Because there are many different tribes in Africa with different languages, a number of countries broadcast in several major tribal languages. Here, too, the specific programming is aimed at the special interests of targeted audiences in order to gain their political support. The difficulty of achieving this is highlighted by the fact that there are some 1200 languages spoken south of the Sahara, excluding dialects, and some 850,000 villages to reach.

In most African countries there is a mixture of news and music on radio and a concentration on entertainment on television. Most countries subscribe to foreign and international news sources, selecting those reports that the controlling party—the government—finds significant and acceptable for its population. The general practice is to combine reports from all news services and organize the selected stories into conglomerate newscasts.

The Mideast and the Arab States

The expansion of individual satellite-receive dishes throughout the region has led to problems with programming in a number of nations. Attitudes toward and censorship of programming varies from Arab state to Arab state. For example, some countries like Saudi Arabia and Qatar are extremely concerned that the content of programs received by their populace does not violate the religious principles that govern their countries.

Two alternative-programming satellite services, largely financed by Saudi Arabian private entrepreneurs, were competing for Arab world audiences in the mid-1990s. MBC, the Middle East Broadcasting Centre, provided a satellite service from London that included western-type programming deemed inappropriate by some governments for their citizens. MBC had a large audience in Saudi Arabia. ART, Arab Radio and Television, provided several satellite channels from Italy, including a children's channel and an Arab movie channel, and also programs showing western lifestyles. Programs from southeast Asia that could be picked up on satellite created similar problems for some Arab governments.

Saudi Arabia's and Iraq's solution in 1994, and Iran's in 1995, was to ban individual satellite receive dishes. Saudi Arabia went a step further. In its efforts to develop television nationally it began to build an extensive cable system, which would be fed programs from satellite that were first screened by the government for acceptability.

In contrast, some other countries are broadening their citizens' perspectives by permitting programming that presents information and lifestyles reflecting the world at large. In the UAE, for example, programs from the west increased in the early 1990s. According to UAE broadcaster Mohammad Abdullah, the government did not interfere with individuals who wished to receive such programming. Abdullah believes that the openness of the government in regard to censorship of such programming and the acceptance of the programs by the public has established a climate that makes it unlikely that programming from other parts of the world will be banned.

Egypt

Egypt has the largest and most influential Arab state system in Africa, providing large amounts of both radio and television programming to other countries. In the early 1990s, however, political differences resulted in cutbacks of programs imported from Egypt. Its various radio networks offer different materials to different populations in Egypt itself. Radio Cairo, for example, is a mixture of music, news, commentary, and other entertainment. "Voice of the Arab" is principally a government propaganda station. Another offering, the Second Program, provides cultural and artistic programming. The Peoples' Program serves special target groups such as farmers, women, youth, and the military, promoting national growth and unity. There is a commercial network leaning heavily on advertising. There is special Palestinian programming. There is even a radio service in Hebrew beamed into Israel.

Egypt's television system provides somewhat the same diversity as does radio, one network oriented to news and education, another to quality cultural and artistic programs, still another with foreign films and programs attractive to tourists—an important factor in Egypt's economy prior to the fundamentalist terrorist campaign in the early 1990s. U.S. soap operas have become popular. However, entertainment programs are not permitted to compete with political and religious programs.

Egypt's program exchange with other mideast countries creates a kind of regional programming, as discussed earlier. While the Gulf War created a schism with Iraq, Egypt continues the programming to and from Jordan, Syria,

Lebanon, and the Gulf States. Wealthy families usually have satellite dishes, even where banned. Depending on the degree of Islamic control, programming reflects or adheres strictly to the standards of Islamic law, as one would find in Syria, for example. Exceptions are international satellite programs in tourist hotels. Nevertheless, the principal percentage of Syrian air time, about 40 percent, is devoted to entertainment. Two categories, news and information, and educational programs, each take up about 20 percent of broadcasting hours. Western programs and program types can be found in a multireligious country such as Jordan. All countries broadcast in Arabic and English, the second language throughout most of Africa and most of the world today.

Countries that were French colonies, those in North Africa and others such as Syria, broadcast also in French. Some, such as Syria, that have undertaken joint projects with Russia, also broadcast in Russian.

While a number of the mideast countries are in Asia, as has been noted previously, they will be included in this discussion on Africa (Figure 5-5). Lebanon, as described earlier, has a privatized system with a western-type diversity of programming. Its programming is probably more eclectic than that of any other nation in the mideast, with the possible exception of Israel, although the latter's government control of the electronic media would appear to make it more difficult to present controversial viewpoints. Although the private stations in Lebanon have great leeway, they are careful not to anger the government lest they lose their broadcasting authorization.

FIGURE 5-5
Satellite TV signals cover the Mideast and Asia. Courtesy STAR TV.

STAR TV Network
AsiaSat 1 Southern Footprint

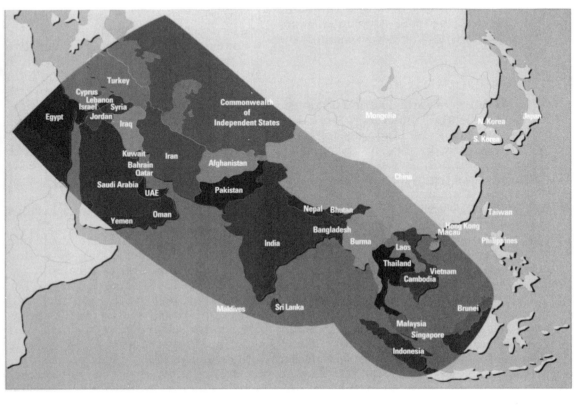

Countries able to receive the STAR TV Network

Afghanistan
Bahrain
Bangladesh
Bhutan
Brunei
Burma
Cambodia
C.I.S.
China
Cyprus
Egypt
India
Indonesia
Iran
Iraq
Israel
Jordan
Kuwait
Laos
Lebanon
Malaysia
Maldives
Nepal
Oman
Pakistan
Qatar
Saudi Arabia
Singapore
Sri Lanka
Syria
Thailand
Turkey
U.A.E.
Vietnam
Yemen

Iraq

Iraq frequently is contrasted with Israel, Lebanon, Jordan, and even Egypt as a tightly state-controlled and censored system. Iraq radio has several different programs, including a main program of news, music, drama and discussion, and another called "Voice of the Masses," used principally for government propaganda purposes. The 1990s Kurdish revolution prompted some programming in Kurdish aimed at presenting the government's view to the rebels. During the reign of President Saddam Hussein, his Ba'ath Party's principles have been presented over the airwaves, all other viewpoints censored and foreign broadcasts jammed, although many Iraqis apparently have been able to pick up signals from other countries. Television, too, is used to present the government's viewpoint, with Iraqi attitudes and lifestyles paramount and western attitudes and lifestyles muted or denigrated. News is controlled by the government and during the Gulf War was censored and frequently distorted, in much the same manner as the U.S. government controlled the Gulf War news reaching the U.S. public.

Kuwait

The country Iraq attacked and occupied, resulting in the Gulf War in 1991, Kuwait, has the same kind of government controlled broadcasting system operated for the same purposes as does Iraq. The Kuwait royal family decides the type and content of programming on radio and television, as well as maintaining complete control over all phases of broadcasting under its Ministry of Information. For some years Kuwait television programming was mostly imported, principally from Egypt, but also from the United States and France. A high percentage of radio listeners tuned in foreign signals, including stations in Saudi Arabia and Baghdad, Radio Monte Carlo, the BBC, and the Voice of America (VOA). Following the Gulf War Kuwait began to establish co-production arrangements with outside organizations and in 1993 was in the process of developing a series on Islamic culture and an Arab version of "Sesame Street."

Iran

In Iran, similarly, all telecommunications is controlled by the government. Following the ouster of the Shah, the new strongly centralized Islamic state named its broadcasting system the Voice and Vision of the Islamic Public of Iran. Its radio networks broadcast in regional languages, one network emphasizing news and information, and the other stressing the arts, such as religious concerts. Over the years underground radio stations have sprung up from time to time, providing usually short-lived counter-government programming. Television programming is similar to that of radio, stressing the religious aspects of Iranian life and presenting the government's points of view. During the pre-Ayatollah regime of the Shah, television carried many western programs, bringing to the people pro-western views of the world.

Some observers believe that negative reactions to these programs helped the Ayatollah to organize public opinion against the Shah and his western-culture tolerances. An interesting side note on the use of communications: in his efforts to create a revolution in Iran, the Ayatollah, from exile, was not able to use the Shah-controlled media; instead, he made thousands of cassette tapes that were smuggled into the country and played in mosques and other gatherings of his supporters.

In the early 1990s television air time was severely limited, and governing beliefs of religion and politics both are strictly applied in programming. For example, women are not permitted to appear on TV (any who do must be properly clothed from head to toe), and English is not permitted.

Nevertheless, most upper- and middle-class families bypassed the state restrictions on their viewing by using satellite dishes to pick up western programming. The influence of these programs on young people is such that in 1994 one young woman, even though knowing she would be arrested, went out of her house dressed like a female announcer she liked to watch on MTV. As noted earlier, in 1995 Iran banned satellite dishes and imposed penalties on any person found to have one.

Israel

Israel has a government system of radio stations and a dual system for television, one government channel and one public channel, although the latter carefully adheres to and is careful to support the government's policies. Dissension is usually in the form of discussions arguing the merits of the policies of the Conservative and Labor parties, and the selection of news items often shows a bias toward the views of one party or the other. One radio service provides entertainment, including music and variety. Another presents news and public affairs. An external service presents programs in various languages, including Arabic, to neighboring nations. Television carries Israeli programs, principally news and public affairs, with a strong emphasis on formal and informal educational programs, produced by the Israel Educational Television Center in Tel Aviv. Satellite and cable, however, have brought many foreign programs into Israeli homes, including some of the sophisticated sitcoms and the not-so-sophisticated evening soaps and crime shows, as well as international news such as that from the BBC and CNN.

Saudi Arabia

Saudi Arabia is a good example of Gulf States broadcast programming. As an Islamic monarchy Saudi Arabia maintains strict political and religious control over all media. All media must conform totally to Islamic law. Radio broadcasts are in Arabic and English. One major service provides acceptable programs of religion, culture, folklore, drama, information, education, and health, all interpreted as appropriate by Islamic scholars and teachers. Another service is composed of more popular programs, but also within the principles of Islamic law. A Holy Koran Broadcasting station programs readings, interviews, lectures, and music dealing with the Koran. Saudi Arabia has tried to establish itself as the focal point of Islamic power and studies in the region and broadcasts its programs in 12 languages in order to reach as wide as possible an audience in Islamic Africa and Asia.

Television, similarly, is used to promote Islamic ideals, using educational, cultural and sometimes entertainment programs to do so. One private station operating in Dahran programs English-language films—but the contents and representation must meet Islamic standards. Because of the many foreign imports on TV, government and religious leaders are especially careful in their censorship of program content. For example, any materials that could be considered sexually stimulating in any way, or materials that depict women in ways not consistent with Islamic practices, such as playing a sport, dancing, dressed in nonacceptable styles (especially Islamic women appearing unveiled), or materials showing gambling or alcohol use are prohibited. In news broadcasts the clergy and the royal family must be properly praised. Other religions are not mentioned.

In Saudi Arabia the state oil company, Aramco, which works closely with U.S. oil interests, funds and operates a few stations oriented to the interests of its foreign employees, many of whom are American.

United Arab Emirates

The United Arab Emirates (UAE) is a microcosm of what some other mideast states would like to do, that is, meld the broadcasting systems of different jurisdictions into one. However, according to a number of key media leaders in the region, as noted in Chapter 1, even if this were possible among a number of Arab countries, it is not likely to accomplish the desired Arab unity. For example, in 1994 Jordan and the Sudan were discussing a joint television production company, but in most situations in the Arab world the political interests are so different—with the exception of general anti-Israel policies—that widespread joint endeavors are not expected.

In the UAE, however, seven federated sheikdoms developed and centralized their broadcasting interests in Abu Dhabi. TV stations broadcast in Arabic and English, and there is an exchange of programs with some other countries in the mideast, particularly Kuwait, Egypt, Iraq, and Jordan. Every type of radio format is used, with local radio programs oriented to the needs and interests of specific regions and communities. The UAE not only broadcasts its signals over a large area of Africa, serving other countries as well, but imports signals, too. However, indicative of the desires of many Arab countries to protect and solidify their own political, cultural and religious bases, and having the resources and organization to do so, UAE produces about 80 percent of its programming.

Other Regions in Africa

Kenya

Kenya is an example of a country that has a substantial radio audience, a concentrated TV audience in its cities, but does not have the monetary resources to produce much of its own programming. Some 90 percent of its TV fare is imported, much of it from the west, especially the United States, with an emphasis on western-type program content. However, as a religious country in which the government attempts to ensure that the media maintain designated family values and promote awareness of national needs and culture, the government influences specific content of the media. It makes certain that nothing considered sexual or violent is presented on TV. Although radio offers popular music and carries news, the music lyrics are carefully watched, and many Kenyans are concerned that the news is too parochial and does not present adequate information about the rest of the world. Because TV is too expensive for most Kenyans, the TV audience is concentrated almost totally in the cities, with virtually none in the rural areas.

Nigeria

With one of the more sophisticated broadcasting systems in Africa, Nigeria operates several radio and TV networks and an extensive external service. Its radio stations principally promote Nigerian culture and national unification, broadcasting in a dozen languages in order to reach most of the people in the country. Local stations serve their individual communities, using some 45 different languages. Considering the fact that there are some 250 different dialects spoken in Nigeria, reaching an entire population in their respective tongues is a prodigious job. Other radio content is similar to that of most other countries: mainly music and talk, with some news and public affairs, and occasional dramas and documentaries. Nigerian radio also serves an educational purpose, providing indigenous peoples in rural areas with informal education and with information relating to their everyday lives, which they wouldn't otherwise receive.

Television programming is largely imported and reflects the LCD fare of most countries. News, as provided by, or supportive of, the government, is an important part of TV fare. A number of U.S. entrepreneurs have attempted to enter the Nigerian media market, but the patterns of corruption have made it difficult for most to do so.

Angola, Chad, Senegal

Some of the smaller countries in Africa still operate their media with a reliance for programming on their former colonizer nations. Angola, for example, a former Portuguese colony, imports about 60% of its TV fare, mainly, as might be expected, from Portugal and Brazil, but additionally from Spain and Cuba. Chad is a former French colony and about 50 percent of its programming, including radio, is in French. Because it is one of the poorest nations in Africa, and has been overrun by Libya, it has few independent resources, and with military control of the airwaves, it does little of its own programming. Almost all is imported and carefully screened. TV fare is about 50 percent entertainment and 25 percent news and information. The entertainment mostly consists of acceptable foreign films: about 25 percent each from France and Italy and about 20 percent from the United States. Radio fare is about 40 percent entertainment and about 15 percent news. Because radio reaches many more people, it is used, as well, for educational purposes, about 25 percent of air time. Senegal, also a former French colony, imports most of its programming from France, but does produce its own news and educational materials.

Zambia

Some countries have used the electronic media as a key educational tool where print illiteracy is very high. One such country is Zambia, which has made excellent progress for an underdeveloped country in using broadcasting. The Zambian Broadcasting Service airs radio programs in English and in seven tribal languages. A government Educational Broadcasting Unit services both radio and television, the latter growing quickly in the 1990s. Television at one time imported most of its programming, but slowly Zambia has begun to produce its own programs, especially in the areas of education, news, and self-help, such as health and agriculture. While entertainment is still not a priority in Zambian broadcasting, as the principal programming reaches more people, more entertainment is permitted.

Zimbabwe

Another country that has used the media to solve the same problems of education is Zimbabwe. It has maintained stronger government control over media content than some countries, but in its efforts to make both radio and TV widespread, it provides varied programming, somewhat similar to the BBC approach. One radio channel is for education, another in various languages to reach outlying groups, a third carries popular shows, especially music, and a fourth is in English with news, information, talk shows, and drama. Television offers a commercial channel, which is dependent on advertising and therefore programs those shows that will attract the most viewers. Its fare, particularly in prime time, consists mainly of U.S. and U.K. entertainment series. The poor economic status of the country prohibits much local production, although one TV channel concentrates on educational programming and has recently been developing other programs of a cultural and artistic nature.

Ethiopia

Ethiopia, also a multilingual country, has used media significantly for educational purposes, too. Radio is the principal vehicle, with both a national network and with individual outlets oriented to different ethnic groups in various languages. With the annexation of Eritrea, Ethiopia also operates an Eritrean radio network. About half of the programs are entertainment, 25 percent is news, and 25 percent is education. Ads are an important source of revenue. TV is also government operated, but with virtually no ads. TV programming is mainly educational, including formal school programs and entertainment consisting mostly of foreign films.

Even small countries with small, basic radio systems are aware of the medium's potential for education. A country such as Malawi uses its scant resources to reach rural peoples with educational and self-help material.

South Africa

A giant in size and, compared to other nations in Africa, in its telecommunications system, is South Africa. The end of apartheid was expected to result in greater freedom for radio and television programming. Broadcasting is both in English and Afrikaans. While specific channels were devoted to programming for the "colored" majority groups in South Africa during apartheid, these outlets were based principally upon political considerations, were controlled by the white minority, and offered virtually no access to the majority groups. In the early 1990s South Africa had nineteen separate radio services serving a variety of audiences. These included a national English-language service, a national Afrikaans-language service, a bilingual national commercial service, a number of services aimed at various sections of the country, such as the Transvaal, the Western Cape, the Natal region, and the Orange Free State, and a youth-oriented bilingual service. There were nine radio services broadcasting in languages of the indigenous majority groups of South Africa, such as Radio Zulu, Radio Swai, and Radio Ndebele. There was also a radio service broadcasting to the country's Indian population. South Africa has an extensive external service, broadcasting worldwide in eleven different languages.

The nation's four television services are principally entertainment oriented, with three of the channels following western formats of news, sports, drama, variety, music, children's, and religious programming. The fourth channel serves special-interest groups and includes cultural and educational materials. About half of the programming is imported, with foreign shows dubbed into English and Afrikaans, and sometimes into the Zulu, Xhosa, and Tswana languages. U.S. programs are among the most popular, but, interestingly, they are not the youth-oriented shows that are top rated in the United States. Instead they are series that appeal to more mature audiences. In late 1993 three U.S. shows were among South Africa's top ten: "Murder She Wrote," "Tropical Heat," and "Columbo."

The symbiotic relationship between political systems and the broadcast media has been demonstrated repeatedly in this book. Just as changes in media technology and use affect the political system and events in a given country, so do changes in a political system affect the media. As dramatic alterations in political philosophies and alignments occurred in the late 1980s and early 1990s in many countries—ranging from eastern Europe to South Africa—so did media developments in the affected nations. South Africa, which converted from minority-controlled racism to majority rule democracy in 1994, is a case in point. For the current status of radio broadcasting in South Africa, see the Appendix, page 152.

ASIA

Programming in Asia seems to reflect the cultural differences among its varied nations to a greater extent than in Africa, perhaps because more Asian countries have the resources to support active television and film industries. In addition, several Asian countries—aside from the mideast, which has been discussed earlier—have become world leaders in trade and have been able to put greater emphasis on self-produced materials for national purposes. Such countries include Japan, China, Taiwan, Hong Kong, South Korea, Singapore, and, by virtue of their location and size, India and Turkey. Other countries are also growing, emerging from third-world status. These nations include Thailand, Laos, Cambodia, and North Korea. As in Arab nations of the mideast, common language creates some regional programming in southeast Asia, in both Mandarin and Cantonese

Chinese among audiences in China, Hong Kong, Taiwan, Singapore, and Malaysia. Joint production and the exchange of news and public affairs programs are done through the Asian and Pacific Broadcasting Union, to which most of the countries in the region belong. Most of the nations in Asia have followed one of two major programming patterns since the introduction of television. In some countries television was an immediate fad, instant entertainment of a kind not available otherwise, for those who had access to TV receivers. In those countries entertainment gradually began to be combined with more serious purposes, such as education and information. In other countries television began as an extension of radio, used to promote the principles and issues favored by the government in power, with entertainment gradually added. In all countries entertainment grew quickly when satellite reception became authorized. In more recent years a number of countries have introduced cable and, with Indonesia as a key example, even pay-TV. In almost every country, however, recent or current tenuous political climates prompted governments to maintain sufficient control over media programming to ensure progovernment content even in light entertainment programs and, at the very least, no antigovernment content.

China

China and Japan are, of course, the giants in Asian broadcasting, but with completely different systems and, accordingly, different programming approaches. The government control of all media in China guarantees that television and radio programming advances the purposes of the state. Program guidelines specify productions that help unify public attitude and action. As designated by the Ministry of Radio and Television, and as described earlier in this book, news, information, and education programs comprise the principal programming in China. Reflecting the strong emphasis on education in the country, Chinese television and radio, on the national, regional, and local levels, provide both formal and informal training, ranging from a Television University, similar to the United Kingdom's Open University, to language courses designed to better prepare the people to be successful in international business co-ventures. English, French, Japanese, and German are key languages taught over the air. As with the British Open University, people can complete credit courses toward degrees through the Television University. It should be remembered that because of China's widespread and voluminous rural poor population, television and radio have been boons to the government's efforts to raise the educational level of its people that it could do in no other way. China actually has more television sets than does the United States, although, because of its size, its per capita ownership is considerably lower. High-quality cultural programming is also a staple of Chinese television, programming designed to promote and unify the cultural heritage of the nation. In the early 1990s, however, as China began to change its economic policies and open up considerably to foreign investment and trade, it also opened up to more entertainment programming on television. China Central Television (CCTV), the government agency in Beijing that supervises all TV in the country, began to import foreign programs, principally from the BBC and from U.S. networks. The favorite genre? Westerns. By 1994 a number of U.S. syndicators had contracts with CCTV for distribution of U.S.-produced sitcoms, soap operas and western and adventure series. From an economic point of view, it is a win-win situation. Because China does not wish to invest too many U.S. dollars, barter is used; this opens up the rapidly expanding China market to U.S. advertisers, making it easy for the syndicators to sell commercial spots for their series.

In interviews in 1992 with CCTV officials Chen Guhua, head of the Program Department, Zhao Yuhui, director of the

Feature Programs Department, and Zhang Jianxin (note that in China family names are always listed first), we learned of the increasing interest in U.S. and other western programming. More important, however, to the future of television programming in China is CCTV's interest in training their writers and producers to understand and to be able to use U.S. formats and entertainment techniques in order to make their own programs of more interest to their audiences while at the same time continuing to present content that reflects and forwards national goals.

While television provides a substantial amount of entertainment in China, radio is oriented mainly to political and social educational purposes. Chinese radio consists principally of news and information, and programs of a cultural nature designed to educate and to stimulate national pride. As noted earlier, public audio speakers wake the people with martial music and news and information at an early hour every morning. Although there are national networks, radio in China is mainly local, serving the informational and educational needs of local entities. Almost every work unit has its own radio station, most of them wired. Otherwise radio programming in China is mostly music, specifically music reflecting national, regional, and local culture.

When Hong Kong is returned to Chinese rule in 1997, drastic changes are expected in its broadcasting system, which has been closer in structure, operation, and programming to those of the United Kingdom than of China. Radio Television Hong Kong (RTHK) has the largest system on the island, and Hong Kong Commercial Broadcasting Company (HKCBC) is almost as big. RTHK offers a diversity of programs, ranging from news and public affairs to light entertainment. Soap operas and dramas are produced locally and are particular favorites of Hong Kong viewers. Commercial stations have similar programming. Both the government and commercial stations emphasize educa-

tional programming, both formal and informal. Radio emphasizes the same content as TV, with about 70 percent of all air time devoted to entertainment, about 15 percent to news, and 5 percent to education. Imported programs come principally from the United Kingdom, the United States, and Japan. Satellite reception has made it possible for relatively affluent Hong Kong to obtain more international programming than most Asian countries.

As the old Kuomintang soldiers are fading away in Taiwan, so is the totalitarian control over media that was established when Chiang Kai-shek seized the country after he escaped from China. In the early 1990s Taiwan was in the process of developing a combination of government and private broadcast services. Both television and radio broadcast in Mandarin and Taiwanese, with some English-speaking programming. Almost all of its programming is locally produced, with less than 10 percent imported, mainly from the United States.

Japan

Japanese on the average are reputed to watch more television than viewers in any other country, with the TV set in the average household on more than eight hours every day. As in the United States, 99 percent of the country's homes have TV sets, and as the nation that is the world's principal producer of TV receivers, including miniature equipment, millions of citizens carry portable TVs as well as portable radios.

Japan's system of private, public, and government networks provides diverse programming designed to suit the tastes or needs of virtually everyone in the country. NHK, as discussed earlier, is the government-supported system comparable to the BBC. It produces programming of high quality, and distributes it on its national network and to other systems throughout the world. In addition, it operates a highly sophisticated program and technology development center, contributing to the advancement of telecommunications worldwide. NHK provides both general programming and designated educational programming, the latter for both children and adults. Japan was the second country in the world, after the United Kingdom, to have a television Open University, with its planning begun as early as 1970 when one of the authors of this book served as a consultant in its initiation and development. NHK's general service is about one-third news, one-third cultural programs, one-fourth quality entertainment, and the rest educational, sports, and other formats. Its programming is determined by a citizens consulting committee, with the aim of maintaining high-quality content, including the promotion of traditional Japanese culture through its drama, discussion programs, news and public affairs, as well as through other programs on the arts.

Commercial stations, on the other hand, carry principally entertainment, with prime time devoted to shows similar to U.S. offerings: sitcoms, crime and adventure, music and variety, and dramas, some historical in nature. A favorite format in Japan is the "samurai drama," which is similar in script as well as historical importance to the American "cowboy drama." Afternoon programming also is similar to that in the United States: magazine formats oriented to housewives. Late-night programming is comparable to the so-called safe-harbor adult programming in the United States, with many of the shows in Japan having erotic themes and contents. Many late-night programs on commercial channels, including interview and game shows, contain nudity, simulated sex, and graphic sexual humor and discussions. In fact, one network, the Nippon News Network, acquired the name "the wild network" some years ago because of some of its late-night programming. One of the most popular soaps in Japan presents simulated sex to a degree not seen in the United States. Violence, nudity, sex, and

graphic language that are commonplace on some commercial stations in Japan would be considered indecent, profane, and obscene in the United States. So-called public stations carry commercially oriented shows, but have more programs of a public affairs, interview, and talk nature than do the so-called commercial stations. Otherwise, they have similar entertainment programming, including an emphasis on sports broadcasts. Virtually all programming in Japan is produced there; very few series are imported, although the various satellite channels originating from the United States and the United Kingdom can be picked up, with their not-made-in-Japan programming.

Radio in Japan is similar to that of most countries in the world today that have extensive television systems. It features mainly popular music and news, but in keeping with the approaches in countries with strong cultural traditions—and contrary to the situation and practice in the United States, for example—Japanese radio includes substantial programming of a cultural and educational nature and occasional quality entertainment such as drama (Figure 5-6).

South Korea

South Korea's is perhaps the next most sophisticated system in this geographic area. At one time a combination of government and privately-owned commercial channels, in the 1980s the government's Korean Broadcasting System (KBS), took control over all the private stations, including majority ownership of the largest private system, the Munwha Broadcasting Company (MBC). Programming reflects national purposes, with one network concentrating on news and public affairs, another on entertainment (but stressing films, drama, and children's shows), and another on more popular fare. Audience preferences are prime-time soaps and dramas, foreign films, comedy shows, audience participation and quiz shows, sports, and

FIGURE 5-6
Information
pertaining to
NHK's various
radio services.
Courtesy NHK.

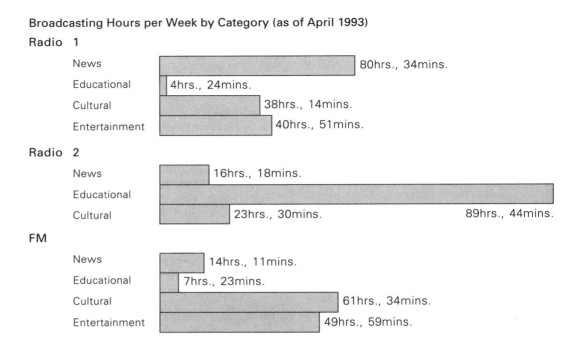

Broadcasting Hours per Week by Category (as of April 1993)

Radio 1

News	80hrs., 34mins.
Educational	4hrs., 24mins.
Cultural	38hrs., 14mins.
Entertainment	40hrs., 51mins.

Radio 2

News	16hrs., 18mins.
Educational	89hrs., 44mins.
Cultural	23hrs., 30mins.

FM

News	14hrs., 11mins.
Educational	7hrs., 23mins.
Cultural	61hrs., 34mins.
Entertainment	49hrs., 59mins.

news. Not much material is imported, with the government establishing limits on the grounds of protecting the quality and integrity of programming; most imports consist of dramas, documentaries, and features from the United States and the United Kingdom. With all of television under either government operation or jurisdiction, material that might be construed as negative to the party in power is not seen.

North Korea

North Korea's system is completely state run, and no foreign programs are permitted without express permission. Both radio and television are used to unify the country's politics and culture. Similar to some of the fare in China, many programs are aimed at worker and family groups. News and public affairs, concerts, drama, and other culture-strengthening programs are prominent on television. Radio concentrates on news, information, and Korean music, with wired systems reaching work units in all areas, similar to the system in China. There is some program exchange with China and with some other coun-

tries whose programs reflect compatible social and economic ideals. Before the velvet revolutions in the USSR, North Korea had a fair amount of program exchange with the Soviet Union.

Thailand

Thailand, the first country in Asia to establish television, developed a combination of educational and government broadcast services, with virtually all the funding coming from the government. All broadcasting is strictly controlled by the government, and most television and radio programming is oriented to supporting the party in power. There are even channels devoted principally to public relations programs about the Thailand Royal Army. Television programming consists of little news other than government propaganda. Entertainment programs, however, play a big role. Thai radio operates similarly, with educational programs approved by the government its principal fare. Radio also broadcasts entertainment, such as music, drama and other cultural content, and some news. Radio broadcasts are in the various languages used in

Thailand, including Thai, English, French, and Vietnamese.

Indonesia

One of Indonesia's most popular channels with those wealthy enough to afford it is a subscription channel that carries mostly foreign programming, including sports, and that broadcasts in both Indonesian and English. Its principal imports are from Australia, Europe and the United States. The main TV broadcasting in Indonesia, however, is by the government-operated Television Republik Indonesia (TRI). About two-thirds of the Indonesian population have access to television, many of them through sets placed in strategic places in small villages and other rural areas. Because of its large land mass and widely dispersed population, Indonesia was one of the first countries to develop a satellite system to reach all parts of the nation.

Programming, under control of the government, is careful to support government policies and to avoid any criticism of the party in power. Official government-prepared programs are also broadcast. Nevertheless, there is a substantial amount of entertainment and educational programming. Because of the economic status of the country and the scattered population, educational programs predominate, especially those dealing with family planning. Because of Indonesia's many different cultures and religions, TV content is careful to try to serve all interests, while at the same time unifying attitudes toward national goals, and to avoid offending any religious group. Strict moral codes are applied to programming, including reverence for the family, marriage vows, and premarital abstinence.

Radio remains the most important mass medium, however, because of its pervasiveness and past impact, the availability of cheap transistor sets, and the necessity to reach millions of people on hundreds of islands over thousands of miles of sea. There are a large number of radio stations throughout the country, some reaching wide areas, some considered amateur stations serving small areas. The principal network is under the control of Radio Republik Indonesia (RRI), which provides the only national news service. Nongovernment stations are prohibited from presenting their own news programs. Entertainment is the principal fare on radio, especially music and occasional productions of drama; educational programs are designed to serve the needs of the scattered, isolated population.

Singapore

Singapore radio and television is operated by the government, which maintains tight control over content. Both radio and television broadcast in the principal languages of the nation: Mandarin, Tamil, Malay, and English. Because of the high costs of production, most television programming is imported: English-language fare comes mainly from the United States, Australia, and the United Kingdom, and other-language fare from China, Taiwan, and Japan. Educational programming is important, conveying the information the government deems important to the people. Radio programming emphasizes music and light entertainment, but includes educational and information programs.

Vietnam

Vietnamese broadcasting exemplifies the use of the media during times of crisis, in this instance the efforts of the new government to unify the north and the south after decades of fighting for independence from foreign powers. This effort was further complicated by the Marxist orientation of the winning north and the anti-socialist policies of the losing south. Because of its need to strengthen national policies and culture, Vietnam's imported programming is principally from countries with similar political views. Programming is produced both for national distribution and by local stations to serve local needs. In

1993, twenty years after the end of the war in Southeast Asia, Vietnam began to open up its borders to foreign visitors and trade, and concomitantly began to relax some of its tight controls on various enterprises, including the media.

Sri Lanka

Sri Lanka is another example of strong government control dictating content that forwards the political purposes of those in power. Radio continues to be the principal medium, as it is in other developing countries with scattered populations. Broadcasting uses the three main languages of the area: Tamil, Sinhala, and English. Music is the leading format, and includes western pop music; however, its content, especially lyrics, is carefully screened to avoid political or social content deemed culturally or morally harmful. In addition to music, program formats include drama, news, interviews and discussion presenting the government viewpoint, and religious programs.

India

Farther north, India approached broadcasting much the same way. After it gained its independence from Britain it used radio as a key tool in unifying the people behind the new government's policies, and has done so ever since. The early 1990s saw a trend toward broader acceptance of outside materials. Radio—All India Radio (AIR)—and television, including the principal network Doordarshan, operate under the control of the Ministry of Information and Broadcasting.

Indigenous music, readings of Indian literature, and speeches by government leaders were and continue to be important parts of programming on radio. Classical and western music, as well as folk music, is also aired, and educational programs, such as agricultural information, continue. Radio carries news bulletins throughout the day, in the early 1990s averaging about 80 national bulletins a day in 10 different

languages, and more than 125 regional bulletins in 64 different languages and dialects of the country.

Television began as an educational tool, using satellite to reach the many isolated villages of India with formal learning and general information not otherwise obtainable. Television penetration to India's cities is shown in the following table:

Cities	STAR TV (,000) Homes	% of TV Homes
Ahmedabad	265	57
Bangalore	214	32
Bombay	946	40
Calcutta	170	11
Delhi	446	27
Hyderabad	86	14
Jaipur	32	12
Kanpur	58	18
Lucknow	41	16
Madras	54	6
Nagpur	53	19
Pune	147	34

Source: Frank Small & Associates; courtesy STAR TV.

Education still comprises a significant part of TV programming, but formats have expanded to include soap operas, sitcoms, quiz and audience participation shows, music and variety, drama, and arts and culture programs. News and documentaries are popular. Drama series unofficially imported from Pakistan get large audiences, especially in the north, where Pakistani-made films are very popular. Because of the religious nature of the country, historical religious dramas are also favorites, the latter keeping viewers glued to their sets week after week. Satellite TV has brought foreign programming to those who have dishes, usually serving a large number of homes wired into one receiving point. CNN, the BBC, and MTV are assiduously watched.

Pakistan

Pakistan's system is similar to India's. Broadcasting is a state monopoly, radio under the Pakistan Broadcasting Corporation and TV under the Pakistan Television Corporation. Broadcasts are mostly in the principal language, Urdu, but include some programming in the other twenty languages widely spoken in the nation. Programming serves the government's aim of unification and supports its policies. While radio imports more than a third of its programming, mainly from the BBC and All India Radio, television imports only about 10 percent of its fare. On radio over 40% of the schedule is devoted to music, with about 15 percent each for news and religious shows, and some 10 percent of the time for commercials. As in India, films and drama, including soap opera and sitcom series, are popular, and comprise over 40 percent of TV air time. News and information shows take up about 25 percent of the television schedule, cultural and educational programs about 15 percent, and sports another 10 percent. Advertising accounts for less than 5 percent of air time on television. Reciprocally, drama series from India are among the most popular TV shows.

Nepal

Nearby Nepal is an example of how economics and geography can combine to limit the development of telecommunications. As a kingdom, its radio programming initially reflected the religious interests and prayers of the royal family members who, like 90 percent of the population, are Hindu, and also featured royal events, ceremonies, and speeches. Not much has changed in recent years, except for the addition of some music, information, and cultural programs. News broadcasts are periodic throughout the day and night. Television initially followed the same pattern: activities of the royal family. However, in the late 1980s TV began to expand to include popular formats, including local soap operas and documentaries, cartoons and science programs imported from Germany, and programs from the BBC. Because of Nepal's terrain, TV is directly receivable only in the Katmandu

Valley, but the government has set up TV receivers in community centers in remote areas that are reached by special relays.

Bhutan

Nepal's neighbor, Bhutan, is an example of an economically and technologically poor country, with primitive, isolated telecommunications conditions compared to most others. In the early 1990s Bhutan had relatively few television sets, and these were tuned to foreign signals until the government ordered TV antennas destroyed, to prevent foreign influences. Radio is totally controlled by the government and reaches relatively few people with its principal fare, government-determined news and information. Because of the mountainous nature of the country, radio signals from nearby nations, including adjacent Nepal, are rarely received clearly.

Turkey

Although geographically in Asia, Turkey frequently is considered a part of the mideast, and closer to Europe in political and social relationships. Turkey's broadcast system, however, developed more akin to those of Asia than Europe. The Turkish Radio and Television Corporation (TRT) for many years monopolized broadcasting. Though organized as a public corporation with a board of directors representing various interest areas, it is government owned and its executives are appointed by the government. In the early 1990s, however, a relaxation of government control in several areas included broadcasting, and permission was granted for competing private channels—not as domestic broadcasters, however, but as imported signals. By 1992 six commercial imported TV channels were competing with six TRT TV channels. The public's preference for entertainment as opposed to TRTs emphasis on educational and cultural programming was immediately evident as TRT was left with only about

25 percent of the viewing audience. The same thing happened to radio, with TRT retaining only about 20 percent of the listening audience.

Films, domestic and foreign, comprise a fair portion of the TRT schedule, with heavy emphasis as well on education, news and information, arts and culture, and programs oriented to special groups in the country. Not surprisingly, the private channels' offerings of sitcoms, soap operas, adventure shows, and other popular fare from United States and European systems have an LCD appeal to more viewers. Programs that feature interview and discussions of issues that previously had not been touched on TRT also have drawn new audiences. TRT radio program genres are similar to those of TRT television. Private radio stations emphasize popular music, including western music that previously was mainly heard privately on record and tape. In the early 1990s Turkey began experiencing the kind of changes in programming that virtually every country in the world has gone through when competitive formats are introduced.

With the increase of satellite distribution in Turkey, TRT decided to use satellites to distribute its programs internationally, and as the 1990s progressed TRT was becoming a major player in international programming.

OCEANIA

Most of the countries in the Pacific Basin are too small or too poor to produce much of their own programming. Those that have television systems import anywhere from 60 to 80 percent of their programs, principally from the largest and most influential country in the region, Australia. Because many of these small countries are developing nations, some with tenuous political stability, they emphasize educational and informational programming in order to unify the people in support of the government and to train them for economic growth.

Australia

Australia has a threefold television system: the Australian Broadcasting Company (ABC), which is a public corporation similar to the BBC, private television channels, and the government-operated Special Broadcasting Service (SBS). All broadcasting is regulated by the Australian Broadcasting Tribunal (ABT), which is under the Ministry of Communications. The ABT has considerable power, much more, for example, than does the FCC in the United States, and it has the authority to censor any given program or program content, including programs from foreign sources.

ABC's programming concentrates on high quality drama, arts and culture, documentaries, public affairs, and news. As many countries do, especially in Europe, ABC has its own symphony orchestras, and frequently offers live concerts. Because ABC is required to be objective in its news coverage, it tends to shy away from controversial stories where its coverage might give the wrong impression. This makes it possible for the competing commercial stations to cover some news areas more effectively than does the ABC.

Television fare on the commercial stations is LCD oriented. Through the 1980s the top-rated programs in Australia were U.S. imports such as "Dallas," "The Cosby Show," and "Taxi." Objection was not so much to the content as to the fact that so many foreign programs, including another U.S. favorite, "60 Minutes," were hurting Australia's TV production industry. An import quota system resulted in a turnaround, and by the early 1990s domestically produced similar formats, especially police series, became among the highest rated. In 1993 Australian TV drama production grew, with an increase of domestic drama on the commercial stations. Part of the growth is due to Australia's role as a major supplier of programs not only to other south Pacific nations, but to much of the rest of the world, including the United Kingdom.

Popular TV program formats in Australia in the mid-1990s included adventure, children's shows, animated shows, mystery, police, sitcoms, and family-romance programs. Among Australia's top-ten-rated shows were reality programs, public affairs, life-style, sitcoms and dramas (such as the U.S. feature "Beverly Hills 90210"), and U.S. feature films. In addition, home shopping and other interactive TV formats were beginning to make inroads.

The Special Broadcast Service, SBS, is designed to serve the needs of the many different ethnic cultures in Australia, ranging from the Aborigines, Pacific Island people, and the growing numbers of immigrants from Asia, the mideast, Latin America, and Europe. Like ABC and the commercial system, SBS operates both radio and TV stations. Its broadcasts are in thirty-seven different languages.

Radio in Australia follows much the same pattern as television. The ABC has separate radio channels, similar to the BBC, one offering popular music, sports, information, news, and other entertainment; another concentrates on cultural programming such as classical music, drama, documentaries, and religious programs; a third serves regional and local stations with a mixture of music and news, with the local stations producing their own programs to meet local needs. Other stations serve the ethnically and culturally diverse populations of the country. Private commercial radio stations, as might be expected, feature local news and light entertainment such as popular music. It is interesting to note that Australia like the United States has two radio systems, FM and AM. Programming, however, is similar to what it was decades ago in the

United States: serious programming, such as classical music and educational programs, are on FM; and light programming, such as pop music and talk shows, are on AM.

As noted earlier, one of Australia's renowned uses of broadcasting is its educational programming, both on radio and TV, to reach children in the isolated areas, such as the Outback, where families' sometimes nomadic existence would otherwise leave the children without any formal education. Alice Springs, in the center of the country, is famous for its School of the Air.

New Zealand

New Zealand's broadcast operations are similar to those of Australia, except that in the late 1980s the New Zealand government began to privatize many of its public utilities, including television and radio, and by the early 1990s the government-supported public entities, Radio New Zealand and Television New Zealand, were being seriously challenged by private commercial stations. Programming is similar to that of Australia: the public TV and radio networks present cultural, educational, information, and public affairs programming, while the commercial channels concentrate on LCD entertainment shows, including miniseries, sports, and Australian films on TV and popular music on radio. As in Australia, locally produced programs are very popular.

Programming in the smaller island countries in the region, as noted above, is limited by the lack of funds for production. With few exceptions, all television fare is imported, mainly from Australia and some from New Zealand. While the lower production costs for radio permit many island systems to create their own radio programming, many use their radio stations primarily as relays for signals imported from Australia and New Zealand.

Appendix to Chapter 5

RADIO IN SOUTH AFRICA: A CLINIC TO HEAL, A SCHOOL TO EMPOWER

William Siemering

Seeing the vast squatter camps in South Africa, it is understandable that more people have radios than sleep on mattresses. Since radio is the dominant medium for the majority of the people, it will play a critical role in meeting the vast educational, health and development needs. One study found that nine out of ten black South Africans have at least one radio and "Three quarters of the population listen to the radio and see radio as the most informative, understandable and entertaining medium."[1] Three-fifths of black South Africans live in rural areas where over 80% have no electricity.[2] Illiteracy rates have varied from two out of five to 63% in rural areas.

After spending two months over the last year meeting with a broad range of people committed to radio, I concluded that South Africa has the potential to develop one of the most diverse and effective radio systems in the world for the following reasons:

- Radio is the least expensive medium, an appropriate technology for this time.
- The high development needs give high motivation for effective use of the medium.
- The majority of the people have a rich oral tradition which is ideally suited to radio.

- Under new leadership, the South African Broadcasting Corporation (SABC) will undergo a complete transformation.
- Radio is personal, imaginative and democratic as it is accessible to all.
- Radio can provide alternative views to the domination of newspapers by a few large corporations.[3]

The SABC is drawing up plans for a public service strand of programming and for the first time independent community stations will be licensed. Community radio can give a voice to the voiceless; be a town hall to build a new sense of community; a school to empower; a stage for the imagination; a clinic to heal; a source of community pride. It can let the leaders know the concerns of the people as a tool for democracy and celebrate the rich diversity of South African cultures in nation building.

The Independent Broadcasting Authority (IBA) created with broad powers to regulate broadcasting is conducting public hearings and will be allocating frequencies for different types of stations: commercial, community, public service and SABC. Dozens of prospective broadcasters are anxiously waiting for the opportunity to file applications and receive licenses. Stations operating with temporary licenses during the transition government, provide examples of the potential for radio in the future.

Radio Zibonele, with equipment costing only R15,000 ($4,453 U.S.), and a 500 watt transmitter, can reach about 20,000 listeners in crowded Khayelitsha near Cape Town each Tuesday morning with two hours of programming. (It operated illegally without a license and

---

[1] *Rural Media* Department of Journalism & Media Studies, Rhodes University, Grahamstown. Don Pinnock, Project Coordinator; p. 31.

---

[2] *The Economist*, April 23, 1994.

---

[3] "Press freedom not cut and dried" by Moegesien Williams, *South*, Cape Town, 17 May 1990. At this time about 93% of all urban dailies and weeklies were controlled by four English and Afrikaans newspaper groups. This has changed somewhat with the Argus group selling its control to Ireland's Independent Newspapers, but there remains domination by a few.

now is off the air pending a new application to the IBA.) Radio Zibonele (which means "we did it together") was started using Dr. Gabriel Urgoiti's experience with community radio in his native Argentina. The elected community health workers, with one week of radio training, conduct the interviews, produce the program and composed the songs and theme music. Dr. Urgoiti believes Radio Zibonele is a tool for primary health care when used *with* community health workers, provides freedom of expression and empowers by demystifying radio. Broadcasting in Xhosa, the health workers/presenters take a holistic approach to health care and therefore not only talk about diseases but water, electricity, education are also related to health. Topics have included tuberculosis, sticky eyes, worms, gastro-enteritis and women and violence. They also remind patients of their appointments at the clinic. The broadcasters receive regular feedback from the listeners when they visit their homes. The station was particularly important during times of violence such as occurred after Chris Hani's death. Dr. Urgoiti wrote, "Radio Zibonele never stopped broadcasting during these times and the main theme was the need for peace in our community." This is an excellent example of grass roots radio growing out of an intimate knowledge of the community it serves.

Radio C Flat, serving the disadvantaged communities of the Cape Flats, was developed by Jeffrey Kleinsmith, who is also the publisher of the community newspaper, *The Peoples Express*. He lists the following role for the station:

- inform, educate and entertain at township level
- train transfer skills and develop talent
- develop, promote and restore pride in local culture and achievement
- empower individuals to make informed decisions
- promote peace, tolerance and mutual co-existence

- and to encourage groups to become constructively involved in local issues, reconciliation, reconstruction, civic, business and development.

In December 1993, the station was granted the first temporary community license for one month and the programming focused on AIDS awareness and education. This spring Radio C Flat broadcast extensive voter education programming. The newspaper reporters rewrite their copy for radio. To facilitate community access, old truck containers are transformed into "community studios" where training and information gathering take place. Mr. Kleinsmith is very entrepreneurial and supports the station through advertising, which is allowed for community stations as long as it is non-profit. It broadcasts in English, Xhosa, Afrikaans and Sotho.

At the other end of the country at the University of Zululand, Robert Ntuli and Gary Mersham of the Communications Department, propose another model. The community station would be based at the university which would construct the studios and provide administrative support. University resources would be used to develop programming on health care, farming, law and small business. The area to be served will include the growing industrial area of Richards Bay and the rural sections where two-thirds of the schools have no electricity.

In Durban, independent producer Julie Frederikse and actor/writer Madoda Ncayiyana produced a series of 20:00 radio soap operas to assist in voter education which were broadcast of Zulu Radio. They were well produced and attracted a large audience. Ms. Frederikse plans to build upon the popularity of this format and identification with the characters to continue the series as part of educating for democracy if funding is secured.

There is much idealism and meritorious goals for community radio as aspiring broadcasters stand before a blank canvas and imagine. Surely whatever appears should be unique to South Africa's history, culture and needs. People new to broadcasting can bring freshness and innovation to the medium.

Inexperience, particularly in management, can also result in failure. While producers/reporters can create their own voice, principles of management and programming from other countries can inform new stations. The creative possibilities of the empty stage of the radio theatre are limitless; but the theatre must be built according to sound architectural and engineering principles or the building will collapse.

Public radio in the United States has evolved into arguably one of the best in the high quality of its news and public affairs, diversity and editorial independence. It is unique in the world in the diverse sources of its funding: listeners, corporations, foundations and government and therefore has much to offer South Africans interested in training. National Public Radio, the primary program producer and distributer, is controlled by the stations (bottom up) unlike the SABC and European state systems which has attraction for South Africans. Because it is listener driven, it is concerned about audience appeal as well as mission.

Regardless of the mission of community radio in SA, it faces four major challenges:

1. **Funding.** Outside donors will undoubtedly need to help launch community radio. Long term sustainability is yet to be resolved. Since the mission of most community stations is to serve the poor and formerly disenfranchised, funding will be difficult from traditional sources of revenue: advertising, corporate underwriting or listeners. An audience demographic of 70% unemployment and no disposable income will not attract advertisers. Local merchants may not be able to afford to advertise or see its value. Accurate audience data may be impossible to obtain in many areas.

   Australia, the Netherlands, Denmark and Germany have supported radio development and now the United States is beginning to do so as well through training and other support.

2. **Defining the Community.** As seen from the examples cited above, there is no simple definition of "community radio." How is community control to be defined? Umlazi Peace Radio near Durban proposes "the involvement of the community in the running of the affairs of their station, its administration, programs, technical, broadcasting, managing, marketing etc." How is this carried out in the operation of the station? Governance should be a means of making the station responsive to the community but not an end in itself or a preoccupation. Just as community radio promotes diversity, so the definition needs to be inclusive and not limited by a narrow orthodoxy.

3. **Limited Number of Trained Broadcasters.** Since the SABC was a tool of the government, tool of apartheid and a monopoly, there was little incentive to learn broadcasting in the past. Eric Louw writes:

   > . . . as of 1992, no significant pool of trained broadcasters or radio administrators really existed in South Africa outside the SABC. Hence the foundation upon which to build successful community radio projects was not in existence.[4]

   Managing a non-profit station is more complex than a commercial one. The sources of funding are more varied, the mission less clearly defined, the board more involved, governance more political and volunteers less experienced. Cov-

---

[4]P. Eric Louw: South African Media Policy. 1993 Anthropos Publishers, p. 77.

ering controversial issues and investigative reporting are risky at most established media; maintaining journalistic ethics and standards under local pressure will require skill and courage.

Training is an immediate and urgent need. It includes workshops, internships abroad, teleconferences, station consultants and training trainers to institutionalize broadcast curricula at technikons and universities. It should be presented in the context of problem solving rather than giving recipes to be followed.

Regrettably, the SABC has not provided a good model of either program production or journalistic excellence.

4. **Program Quality to Meet Competition.** Given these limitations, how will community radio produce programs of distinction and sufficient quantity to compete with the present most popular Zulu Radio and Metro and the new commercial stations? Will the programming be so appealing that listeners will tune away from their present favorite station for a new one? Will local, familiar voices be enough? To succeed community stations will need to focus on the three "p's" of broadcasting: programming, power and promotion. Commercial broadcasters can argue, "Because we have the most listeners, we are truly serving the community."

Much of the education/development programming is based on what is assumed others *need* to hear. The challenge is to produce programs so people will *want* to listen to what they *need* to hear. There are few models of engaging successful educational radio and South Africans may need to be the most innovative in this area.

Radio is a companion. If it isn't there when most people use radio, they won't seek it out. Most stations will find it difficult to fill 10–12 hours of programming a day. Most underestimate the staff necessary to sustain a full service station. South Africa could leapfrog technologically and using Internet interconnect the community stations to not only share programs *within* the country, but open up exchanges with stations abroad as well.

The Open Society Foundation for South Africa (OSF-SA) was founded in 1993 by George Soros with an initial grant of $15 million over three years to promote the values, institutions and practices of a civil, democratic society. It has established four priority areas: (a) education for local government; (b) youth development; (c) rural community development with a focus on empowerment of women; and (d) radio. OSF-SA has allocated over R2,722,000 ($800,000 US) for the development of radio through grants for planning, equipment, training/internships and programs production. This is the most comprehensive support for radio in SA. OSF-SA is looking for partners to share in the support of this work.

When I spoke with eager broadcasters, they frequently spoke of radio as a healing medium, bringing together people who have been forced apart, creating in the ether, a community that does not yet exist on the ground. For example, compelling programming could include oral histories of the last forty years, or take multiple perspectives of a universal theme such as growing up which would be inclusive of all South Africans. Developed as co-productions, they could be engaging in both South Africa and abroad. This would be training of the best kind, resulting in programming excellence.

Radio *can* be central to reconstruction and development but it requires an openness from all to innovation, willingness for South Africans to learn from their American colleagues and for Americans to listen to the rich stories and remarkable transformation of their South African hosts. It should sound like no other radio ever produced. New radio can help build a new South Africa.

6 Freedom and World Broadcasting

As indicated in the overview of broadcast systems in Chapter 2, all countries in the world restrict the freedom of their respective telecommunications systems for what each country considers just and moral purposes. Although most countries pay lip-service to the principles of freedom of information and broadcasting as stated by the United Nations and UNESCO, their practices are usually something else. Some countries establish constitutional requirements to promote a given ideology: political, social, economic, or religious. China, for example, uses the media to solidify national support of its economic and political system; many mideast countries use the media to strengthen their Islamic principles and controls. Some countries do not have specific constitutional or legal restrictions, but censor or dictate to the media when they deem it necessary to achieve a given national purpose. For instance, both Iraq and the United States successfully controlled and censored their media's coverage of the 1991 Gulf War in order to maintain the support of a possibly rebellious public for their respective war policies.

Media freedoms almost always are different in word than in deed. Almost all countries, even those considered to have intensive restrictive control over their telecommunications systems, have some degree of constitutional guarantees of freedom of the media. Some enumerate those guarantees; others state them in broad terms along with reservations relating to the country's na-

tional security, morals, or ethics; others qualify their freedom guarantees by subordinating them to other national laws; and still others with media freedoms list specific exceptions that tend to vitiate the basic guarantees. Nowhere are media freedoms absolute. Even in the most democratic countries, such as the United States, there are restrictions, usually stemming from nonmedia legislation such as obscenity and libel/slander laws, and the government-in-power's interpretation of national security requirements.

Not only are governments responsible for restrictions on media freedoms, but media practitioners themselves are often complicit in both external and internal censorship. Despite many codes relating to the freedom of the press and media, which proliferated after the World War II defeat of blatant dictatorships, many journalists and others in the media who developed and subscribed to such codes were among the first to break them when either the government, the media owners, the advertisers, or the public exerted any pressure. Practitioner and journalist self-developed codes are, of course, voluntary, and there are no mechanisms for enforcing them (Figure 6-1).

In the United States, for example, complete disregard for the guarantees of freedom of speech and thought in both the Constitution and in voluntary codes was the hallmark of the McCarthy era of the 1950s. Media executives in every network and almost every station in the country blacklisted performers, writers, directors, and others solely on the basis of even a passing implication of communist sympathies. No court indictment or proof of any kind was required; as in Nazi Germany, it was enough for a person to be accused to be considered guilty of being a traitor. Virtually no media leaders, with the exception of Edward R. Murrow and Fred Friendly of CBS and precious few others, had the integrity to stand up against the blatant censorship and purging of ideas and people to which right-wing political groups or individuals objected. U.S. constitutional guarantees of media freedoms were completely ignored. Given this behavior by the United States, one of the world's leading democracies, it is no surprise, therefore, to find that many other countries' constitutions also guarantee media freedoms, but contain caveats that result in just the opposite in practice.

Commitment to media freedoms depends largely on the political system operative at any given time in any given country, and the ethical history and attitudes of media practitioners in that country—rather than on the legal guarantees. For example, in Denmark and the Netherlands there are no journalistic codes. Yet, it is generally perceived that media practitioners in these countries have acted with greater ethical protection of media freedoms than practitioners in most other countries. Going beyond simply protectiing media freedoms, media practitioners can conscientiously stimulate attitudes and practices of freedom of speech and the press. This happens in very few countries. Denmark is an example in which it does happen, where the Danish Press Council acts boldly and consistently to maintain and forward such freedom.

Finally, many countries' constitutional provisions and principles, especially in small and developing nations, did not emerge out of the individual country's own ethical background, but were either borrowed from larger countries that provided political and/or fi-

FIGURE 6-1
Cyprus's public broadcasting corporation declares its mission.
Courtesy PIK.

MISSION STATEMENT

«*T*o become an efficient, competitive broadcasting organisation capable of operating independently and profitably in the free market, whilst preserving its identity as Cyprus' public broadcasting corporation and promoting Cypriot culture and national identity».

nancial assistance in the development of telecommunications systems and policies, or were carried over from the principles imposed by former colonial powers. Two examples are India, which retains the basic concepts of media freedoms and controls of its former ruler, the United Kingdom, and Japan, whose post–World War II constitutional codes were either imposed or strongly influenced by the occupying power, the United States.

The age of global communication satellites suggests that the need for global agreement on media freedoms is becoming more and more important. Yet, in the mid-1990s, despite growing international telecommunications interconnections, exchange, and joint projects, the political, economic, and social diversity among all the world's nations suggests that a general global code of freedom of speech and press is not likely to come about in the foreseeable future.

THE AMERICAS

United States

With the notable exception of the United States, constitutional guarantees usually are stated in positive terms: for example, "there shall be freedom of the press." The United States, in the First Amendment of its Constitution, puts its guarantee in negative terms: "Congress shall make no law . . . abridging freedom of speech, or of the press." (Inasmuch as the writers of this book are more familiar with U.S. principles and practices than with those of any other country, the United States is used as the main basis for comparison with other nations and regions of the world. For example, when referring to media freedoms this book will occasionally use the term "First Amendment rights," although the country being discussed may not necessarily have either a constitution or a freedom of speech or press amendment.) Ironically, the U.S. negative statement has

resulted in more positive application than in virtually any other country, including those where the guarantee is stated in positive language. The prohibition on restrictions on the media and on speech has dissuaded legislators from changing such freedoms and have persuaded jurists to uphold them.

The United States, in practice, comes close in its media freedoms to the United Nations Universal Declaration of Human Rights, article 19 of which states

Everyone has the right to freedom of opinion and expression, this right includes freedom to hold opinions without interference and to seek, receive and impart information and ideas through any media regardless of frontiers.

The United States' Radio Act of 1927, which established the Federal Radio Commission, and the Communications Act of 1934, as amended, which established the Federal Communications Commission, provide legal authority for government regulation of telecommunications. However, regulation is not necessarily equivalent to control. The Communications Act, while giving the FCC authority to prepare and enforce rules of governance and operation of the media in the public interest, also forbids the FCC from censoring any media content. At the same time, however, the act also authorizes the FCC to take action against any licensee that violates the act's ban on indecency, obscenity, and profanity. From time to time rules, regulations, and court decisions established policies such as the Fairness Doctrine, which required stations in certain circumstances to provide an opportunity for more than one side of controversial issue to be heard, and Ascertainment of Community Needs, which required stations to present programs dealing with key issues in their communities. But at no time has the U.S. government been permitted to precensor any given program. In fact, the FCC may not even comment on any program material, even if voluntarily submitted for clearance by a licensee. In the 1990s the FCC's imposition of time restrictions and placement for so-called indecent programming began to be reversed by the federal courts. Because of the First Amendment there can be no law permitting the government to stop any medium from disseminating information. Unlike most countries, there is no law that even permits the President of the United States to seize the media, even in times of war or crises. In other words, contrary to most other countries, the media in the United States are not legally subject to any prior restraint. In fact, however, as in most other countries, the U.S. government does exercise prior restraint and censorship when it does not want the public to learn the truth about certain controversial actions, such as in the Gulf War and the invasions of Panama and Grenada.

Canada

Canada provides protections for its telecommunications systems similar to those in the United States. Patterned after the British system, Canadian media have no prior restraints and, in many instances, are able to broadcast content that in the United States might result in after-the-fact punitive measures. Canada's voluntary practitioner codes include that of the Canadian Association of Broadcasters, which, like the now-abolished codes of the U.S. National Association of Broadcasters, carries little enforcement teeth and can be ignored at will, as it was in the U.S. Similarly, the code of the Radio and Television News Directors Association (RTNDA) may function more effectively, but relies on voluntary adherence as does its U.S. counterpart. The RTNDA has done little to prevent unethical conduct or abuses of media freedoms by its members.

Latin America

Latin America—that is, Central and South American countries—is different. Most of these countries are nondemocra-

cies or emerging democracies, with little history of media freedom. The media in some of these countries are able to escape total oppression by their governments through the protection of the church; but, of course, that means the media must represent the doctrines of the church or at the very least refrain from any criticism of the church. Most Latin American countries' constitutions have some provisions for restrictions on media freedoms, specifically articles that permit prior restraint in one or more designated content areas. Dictatorships and democracies ruled by the military by and large maintain absolute control over the media, with their content necessarily promoting the policies and leaders of the government in power.

Mexico

Key examples of democracies closer in proximation to the U.S. concept than most other countries in the area are Mexico and Brazil. Mexico's 1960 Federal Law of Radio and Television ostensibly implements the Mexican constitution's declaration of freedom of speech and the press. However, as with many other countries, its promulgations on the cultural and social contributions of the media, laudable as they may be, also create restrictions on freedoms. For example, the 1960 law's requirement that the media promote Mexico's culture and traditions resulted in the media being required to carry at least thirty minutes of programming each day praising Mexican family life and national unity, these programs produced by the government itself.

Concomitantly, the 1960 law specifically prohibits the broadcasting of any program that might "corrupt the language and is contrary to the accepted customs of Mexican culture."

Brazil

In Brazil, broadcast laws similarly contradict constitutional guarantees of free speech. A National Security Law and a Press Law have permitted the military, over the years, to censor any broadcast material that might be construed as threatening the national security. Even though Brazil's government officially moved from a military dictatorship to a democratic system in 1985, the laws governing telecommunications have remained largely unchanged; however, since that time, the more onerous provisions of the laws have rarely been enforced. Other democracies in Central and South America have followed comparable paths. It is important to note that the growth of international communications and the rapidly approaching global telecommunications village are affecting the degrees of freedoms and censorship. In part, censorship becomes difficult when a large percentage of programming comes from foreign countries (estimated as high as 90 percent for television in Mexico). In part, growing exportation as well as importation of programming tends to impose a general standard of decreasing censorship on all participating countries.

Chile

Nondemocratic countries in the region appear to emulate each other's telecommunications policies. When the United States engineered the overthrow of the democratic, but socialist, government of President Salvador Allende in Chile in 1973 and installed a right-wing dictator, General Pinochet, all media became tightly controlled by the government, and were used solely for government propaganda purposes. Even privately owned stations were forced to participate in government information dissemination. Although Chile's 1925 constitution provided for speech and press freedoms, and a revised 1980 constitution also has a provision for freedom of expression, prior restraint—that is, censorship of all material before dissemination—is permitted under other constitutional clauses. As might be expected in a military regime, slander of the armed forces and of government officials is forbidden and severely punishable, with truth not considered a satisfactory defense. As have many other countries, Chile introduced window dressing for international face-saving purposes. The antidemocratic regime in 1973 issued a Chilean Code of Conduct for Broadcasters that ostensibly established basic freedoms for the media. However, the code was written by the Chilean National Television Council, under strict control of the government, and in fact strengthened the government's control over the media. The election of the more moderate President Aylewin in 1990 suggested changes toward greater freedom in Chile. However, during the first few years of the new government that labeled itself as democratic, no significant changes had been made in telecommunications policy.

Peru

When Peru's 1979 constitution restored constitutional democracy to that country, it included the guarantee of freedom of expression, and "the right to information, opinion, expression, and dissemination of thought, by word, image, or print, and without previous authorization, censorship or any impediment, subject only to legal liabilities." As in most countries, it is the reservation beginning with "subject to," "with respect to," "with the exception of," or a similar phrase that negates the ostensible democratic guarantees. For example, the Peruvian constitution allows the government to suspend various rights, such as freedom of speech and press, whenever it declares a state of emergency. Nevertheless, for some years the media were relatively free. But in 1992 Peru's newly elected president, Alberto Fujimori, suspended the constitution and enacted new laws that made all media subject to complete content control by the government.

Other Latin America Countries

In Ecuador the new constitution enacted with the return of civilian rule from military dictatorship in 1979 guarantees freedom of communication, as well as freedom of speech, press, and religion. In fact, it even provides for penalties for anyone who attempts to obstruct freedom of speech and information. However, its new president in 1984, Leon Febres, began to use the media for political purposes, instituting control of content and even military takeovers of the media. This is the pattern in many other nondemocratic countries in the region, as well, where the volatile political situations may one day establish principles of freedom of information with greater or lesser degrees of implementation, and the next day find those rights usurped by a presidential or military dictatorship. Uruguay moved from a military government to a constitutional government in 1985, but its media remained under control of the military. Paraguay's constitution guarantees "freedom of expression and information and freedom of the press." But Paraguay, a dictatorship since 1954, absolutely controls all media under the president's powers to declare a state of siege—which has been in continuous effect since the 1960s. Even where there is no constitutional reservation about freedom of expression and information, as in Honduras, there is still a means of government control of the media. Indeed, the Honduran constitution forbids the closing or takeover of a telecommunications outlet because of the dissemination of any ideas or thoughts. However, all journalists must, by law, belong to a government-controlled press organization; this requirement of course gives the government power over who may function as a journalist, especially those from foreign countries whose stories the government may not like. Beyond that, the government has been reported to have threatened violence and even death to journalists who do not follow the government line. Further, through its licensing policies the government may deny a license to any station it disapproves of. Aside from the law, however, the government controls the media through the close ties of the owners of the most powerful media to the military and to government leaders. The media, therefore, by and large operate with little freedom despite strong constitutional guarantees.

While El Salvador ostensibly replaced its military regime with a constitutionally elected president, the military has remained in control, as evidenced by the lack of any action against the death squads that accounted for murders of not only political dissidents but of foreigners, including many Americans, who objected to the government policies of oppression. While the Salvadoran constitution does not provide for the seizure of stations, the military has easily controlled media content by threats against any station or individual who might be critical of the government.

Guyana has no laws restricting freedom of speech, but because broadcast stations are controlled by the government, there have not been any effective media freedoms.

There does seem to be a correlation between economic and democratic political stability and freedom of the media. While the two principal democracies in the Americas, the United States and Canada, put the least constraints on the media, the more volatile the political system and the more difficult its economic circumstances, the greater the government's need to control the media seems to be.

EUROPE

While constitutional protections in many European countries approximate those of the United States, many European guarantees of freedom of speech and press include explicit reservations, some stated directly, others contained in different legislation. By and large there is less prior restraint or censorship in western Europe than in virtually all other nations. In 1990 the member countries of the Council of Europe agreed on principles for cross-border transmission of television signals, including the following:

> The Parties shall ensure freedom of expression and information in accordance with Article 10 of the Convention for the Protection of Human Rights and Fundamental Freedoms and they shall guarantee freedom of reception and shall not restrict the retransmission on their territories of programme services which comply with the terms of this Convention.

United Kingdom

In the United Kingdom prior restraint is rarely exercised; one notable exception was the Parliament's threat to reduce the BBC's funding if it carried a program on the Northern Ireland conflict that contained material the government believed favorable to the Irish Republican Army. A compromise was reached whereby the program was carried, but specified segments were eliminated. Obscenity laws in the United Kingdom do create voluntary prior restraint on the part of broadcasters. Further, contempt of court charges can be brought against the media on the grounds of hampering a criminal investigation, and national security principles can and have been used to prevent the showing, and even allow the seizing, of broadcast materials. As in the United States, such laws create a prior self-restraint, a chilling effect.

Western Europe

France

While French law also proclaims the freedom of speech and press and the peoples' right to know, it has always placed conditions on such freedoms. Principally, a determination of national security has consistently allowed the government bureaucracy to restrict the availability of information relating to government operations. Coupled with this are the laws of libel, which can be used against the media even when the person involved is a public figure. On the other hand, France has one of the oldest provisions allowing the right of reply to a personal attack.

Italy

Italy's 19th century "Statuto" on press freedom is illustrative of the conditional guarantees that mark most democratic countries: "The press shall be free, but the law may suppress abuses of this freedom." However, the growth of telecommunications coincided with the establishment of Mussolini's fascist dictatorship in the 1920s, and radio was developed as an arm of the government, subject to government censorship. In fact, Mussolini was once quoted as saying that if it weren't for radio he would not have been able to achieve the control he had over the Italian people. Italy's post–World War II constitution includes a section on freedom of communication. Article 21 states

> All are entitled freely to express their thoughts by word of mouth, writing, and by all other means of communication. The press may not be subjected to any authority or censorship. Restraint is allowed only by order of the judicial authorities, for which motives must be given, in the case of violation of the provisions which the said law prescribesfor identifying responsible parties. . . .

In 1975, following a 1974 court decision guaranteeing the right to information for every citizen, Italy passed a law guaranteeing the independence of RAI, the state radio and TV system, and the following year opened up the spectrum to private broadcasters.

Spain

Spain followed a similar course, establishing a state radio system in 1923, which came under strict and total government control with the ascendance of dictator Francisco Franco in 1937. In 1977, after Franco's death and the establishment of a constitutional monarchy, broadcasting reforms included proclaimed freedom of speech and, shortly thereafter, the authorization of private stations. In Spain's constitutional equivalent, "Basic Rights and Public Liberties," freedom of communications is addressed. Article 20 lists the right to "express and disseminate thoughts freely through words, writing or any written means of reproduction . . . to communicate or receive freely truthful information through any means of dissemination . . . the exercise of these rights cannot be restricted through any type of prior censorship." As in the United States, the Spanish constitution precluded prior restraint or prior censorship. Nevertheless, the article also states that "these liberties have their limits in respect for the rights recognized . . . in the precepts of the laws which develop [them], and especially in the right, honor, privacy, personal identity and protection of youth and childhood." One might ask, as well, who determines what is "truthful" or "untruthful" information?

Portugal

Spain's neighbor, Portugal, also went through the transition from a long-term dictatorship to a democratic state. Its new 1989 constitution has separate articles referring to "freedom of expression and information" and "freedom of the press and mass media." While it guarantees freedom of the press, its verbiage, like that of documents in many other democracies, leaves room for exception: "freedom of expression and creativeness for journalists and literary collaborators as well as a role for the former in giving editorial direction to the concerned mass media, save where the latter belong to the State or have a doctrinal or denomination character. . . ."

Greece

Another example of a European country that has undergone governmental shifts from democracy to military rule to democracy again is Greece. When it developed a new constitution in 1975 after throwing off a dictatorship, it added a "freedoms" clause: "radio and television shall be under the immediate control of the state and their purpose shall be the objective, and on equal terms, transmission of news and information as well as works of literature and arts." The determination of what is objective is, of course, up to the controlling party. The prime minister of Greece has powers similar to the heads of state in other democratic countries in regard to the media: to determine programming when deemed advisable for national security purposes or in a national emergency.

Scandinavia

The area in Europe with perhaps the greatest freedoms of press and speech is Scandinavia. Traditions precluding censorship of virtually any material extend back two and more centuries. In Denmark, for example, the media may present materials which in many other countries might be considered pornographic. Freedom of information laws are strongly enforced in Scandinavia. Sweden's law goes back to 1766 and gives any citizen access to any government file, with the exception of documents relating to national security and foreign policy. Unlike the U.S Freedom of Information Act, which entails considerably more restrictions, files must be made immediately available without requiring the requester's identification or reasons. This gives Swedish journalists greater freedom to access as well as publish material than journalists have in most other countries in the world. Denmark, on the other hand, has many of the restrictions of other countries: internal working documents, foreign policy documents, memoranda between government ministers, certain kinds of statistical data, documents relating to crime investigations, materials relating to public order, and papers concerning public or private economic matters.

Except for national security purposes, the media in Scandinavia generally have excellent freedoms, including the reporter's privilege not to divulge sources.

As far back as 1766 the Swedish parliament passed what is generally considered the first modern press freedom act, giving the press the right to act as a public watchdog. In fact, unlike most countries and those states in the United States without reporter shield laws, the press in Sweden may not disclose confidential sources. Exceptions to this protection are occasionally made in criminal cases. Similarly, in Denmark the media are empowered by law to protect their sources, except where a court may require disclosure to prevent someone from going to jail. However, in a 1986 Danish court ruling this press freedom was not deemed applicable to the broadcast media. Nevertheless, in practice the broadcast media have consistently been given this privilege.

Almost every country that has a freedom of speech and press guarantee also has a codicil listing libel or slander as an exception. (Libel is written or printed material falsely ridiculing or injuring a person's reputation; slander involves false statements spoken or aired, as on television or radio.) In the United States the laws make it very difficult to prove libel: the report must be proven untrue, it must clearly refer to the aggrieved party, it must have been disseminated publicly, and it must have caused harm to the party claiming libel. Further, public figures have to meet a stricter standard and prove not only neglect on the part of the alleged libeller, but actual malice or reckless disregard for the truth, plus monetary as well as emotional harm. Freedom from libel in Sweden excludes pornography, insults to the chief of state of a foreign country, and clear defamation of religious liberties. Denmark does not protect the media from libel when they violate the personal honor of another by offensive words or acts or by making or spreading accusations likely to disparage a person in the esteem of the other citizens. However, the media may use "justified protection of obvious public interest" as a defense against libel. While Denmark does not make legal distinction, as does the United States, between public and private figures in libel cases, politicians nevertheless may expect to be criticized by the media to an extent that might be considered libelous by a private person. Norway's objectivity and fairness in the media is principally voluntary, through the code of the Norwegian Broadcasting Association. As with all such voluntary codes, there is no mechanism for punishing violators.

In the Netherlands the government guarantees freedom of speech, but emphasizes the need for broadcasters to provide programs that serve the needs and enhance the welfare of the people. Germany, reacting to the complete abolition of freedom by its Third Reich, now has strong written guarantees for freedom of speech and the press. However, the regional nature of broadcasting in Germany, with citizen councils serving as directors of the systems, tends to reflect the political attitudes of the given region in programming decisions. Cable, as a private enterprise, similarly reflects the attitudes of the owners of the cable systems.

Eastern Europe

As previously stated, eastern Europe is still an anomaly as this is being written. While the velvet revolutions of the early 1990s established new press freedoms and new constitutional guarantees in most eastern Europe countries, the vicissitudes of governments and their inability to move slowly toward democracy, as opposed to attempts at instant capitalism, have created crisis situations that have prompted even so-called democratic rulers, such as Boris Yeltsin in Russia, to seize totalitarian control of the media. In principle, the restructuring of the former Soviet Union and Soviet bloc governments has followed the political and economic tenets of the United States, and the U.S. and British Commonwealth approach to telecommunications control and operations.

However, without a slow, prepared transition, in most instances chaos has resulted, with telecommunications policies in limbo. One of the difficulties for eastern European countries is the concept of private ownership, the staple of U.S. broadcasting but totally new and, in many cases, incomprehensible to the new democracies. Therefore, the retention of telecommunications media by government results in a concomitant government control and, at the very least, strong influence on content. Further, the words in a communications law or the statements in a constitution, laudable out of context, may nevertheless be empty in application. Elena Androunas, formerly a professor of mass communication at Moscow State University and currently a communications consultant in Russia, notes that a frequent comment in Russia regarding press freedoms is that any writer or editor can publish anything she or he likes—but only once in a lifetime. Even in eastern European systems where there is now relative freedom compared to recent restrictions, media professionals cannot forget the old wariness and are especially careful about what they write or air. Further, the bureaucratic structures of the old system still remain in many instances, building in a restrictive atmosphere. Broadcast officials are those appointed, or at least approved, by the party in power, and those persons' viewpoints are rarely challenged.

The issue of economics cannot be overemphasized. Because media are so expensive to operate, it is necessary for the government to either operate or subsidize the media in many of these countries. By the very nature of its role, government is then in a position that it cannot overlook: use of the media for its own purposes. Finally, because they know how powerful the media are in influencing peoples' minds and even actions, the emerging democratic governments are afraid that uncontrolled media may shatter the still-fragile threads of democracy.

Bulgaria

Bulgaria is a case in point. Its constitutional guarantees of freedom of speech and press go back to its constitution of 1879: "The press is free. No censorship is permitted." Under its Soviet influence, the constitution of 1971 reads: "Citizens enjoy freedom of speech, press, meeting, associations, and demonstrations." And its new 1991 constitution as a republic states: "The press and other information media are free and not subject to censorship." But then it adds: ". . . information may be suppressed or confiscated only through an act of the judicial authorities, when good mores are violated or if it appeals for a violent change in the constitutionally established order, the commission of a crime, or an act of violence against an individual." Therefore, the new, ostensibly democratic Bulgaria echoes essentially the same restrictions against the media that the previous nondemocratic regimes did. Perceived sedition also invites suppression of the media, comparable to the United States actions through the Smith Act and the McCarthyism of the "cold-war" era. Further, broadcast channels are run by political parties and the news almost invariably reflects the point of view of a given political party.

The Czech Republic and Slovakia

Czechoslovakia had one of the freest media operations as a member of the Soviet bloc, with control centered in a Federal Office for Press and Information, but relying heavily on Communist Party loyalties of the media operators to self-censor any questionable material. With the split in 1992 into two countries, the Czech Republic and Slovakia, two individual constitutions were developed, both of which guarantee freedom of speech and information without censorship or government control. Both countries, however, assert the duty of the government to protect the rights and freedoms of individuals and the security of the state, citizen morality, and the public health. Although moving towards a market economy, the Slovakia constitution notes that "business activities in radio and television broadcasting may be subject to state authorization." Both republics, however, assert the government's responsibility to provide freedom of information access for the media.

Hungary

Hungary has long had one of the strongest constitutional guarantees of media freedom, even through its years in the Soviet bloc. Its 1949 constitution states that: "everyone has the right to freedom of opinion, as well as the right to receive and disseminate information of public interest . . . the Republic of Hungary recognizes and protects freedom of the press." This was, however, not practiced. In 1990 freedom to broadcast was officially extended to everyone. However, as this is being written the broadcast media still remain under strict government control, with no private stations, and all telecommunications personnel approved by government-appointed directors. In 1993 there was pressure for change, as thousands of students demonstrated in Budapest for freedom of the press, calling continued state control of radio and television a danger to the fledgling democracy.

Romania

Romania's new constitution, developed after the country deposed its dictator, Nicolai Ceaucescu, is remarkably current and extensive in promising media freedoms:

the freedom to express ideas, opinions, and beliefs, and the freedom of creation of any kind—orally, in writing, through images, by means of sound, or by any other means of public communications—are inviolable. Censorship of any kind is prohibited.

But, in keeping with the worldwide practice, caveats are added:

The law can compel the mass media organs to make public their sources of financing.

Freedom of expression cannot be used to damage the dignity, honor, or private life of an individual or his right to his own image. The law prohibits defamation of the country and the nation; provocation to war or aggression, and to ethnic, racial, class, or religious hatred; incitement to discrimination, territorial separatism, or public violence; and obscene acts, contrary to good morals.

The constitution, to further spur the principles of democracy, states that:

The public and private mass media organs must ensure that public opinion receives correct information. The public services of radio and television are autonomous. They must guarantee that significant social and political groups have the right to broadcast.

In addition, individual radio and TV station owners, producers, and writers may be held directly liable for violations of the constitutional provisions.

While the restrictions may be considered necessary and proper for the successful growth of democracy, nevertheless the question must be asked whether in a democracy freedom of speech and press should be absolute, and whether these constitutional requirements in fact curtail media freedoms.

Poland

Poland exemplifies the ship of state that seemingly has no rudder when it comes to telecommunications. Throughout its history Poland has been either under the control of other countries, or partitioned, or added to by other countries, and has largely been unable to govern itself efficiently even when functioning as an independent political entity. Its new constitution of 1989, as it achieved independence from the Soviet bloc, proclaimed press freedoms, but in language remarkably similar to the language already in the Soviet constitution:

Poland shall guarantee its citizens freedom of speech, of the press, of assembly and gatherings, of processions and demonstrations. To put these freedoms into effect, the . . . people . . . shall be given the use of printing shops, stocks of paper, public buildings and halls, means of communication, the radio, and other necessary material means.

Yet, the media continue to be run by the government, although, as this is written, privatization of media has begun. While the government itself has relinquished strict control over content, the reversion against the atheistic principles of communism has given the Catholic church strong power in parliament as well as among the people. The media cater to the whims of the church in Poland, both in coverage and in content, in effect substituting a degree of religious control for political control.

Estonia, Latvia, and Lithuania

It would seem as though the Baltic states, Estonia, Latvia, and Lithuania, lumped together by the western press, would have similar policies toward media freedoms. They are as alike and diverse as the other eastern Europe countries as a whole. While the Estonian constitution of 1920 and that of Lithuania in 1922 had guarantees of freedom of speech and press, albeit both contained restrictions in regard to morality and state security, the Latvian constitution of 1922 made no mention of press freedoms. In 1978 a new Estonian constitution has lengthy guarantees for free communications, specifying that such communications strengthen and develop the state's socialist system. At this writing both Estonia and Latvia are in the process of writing new constitutions, which presumably will include media freedom guarantees comparable to those in the west that are being adopted by other eastern European countries. Lithuania wrote into its new provisional constitution in 1990 far-reaching guarantees of speech and press freedoms:

Citizens of Lithuania are guaranteed freedom of expression, press, assembly,

street demonstrations and rallies. The implementation of these freedoms is guaranteed by allowing working people and their organizations to use public buildings, streets and squares, by the broad dissemination of information and by according them access to press, television, and radio. These political freedoms cannot be used to promote racial and national enmity and antihumanitarian views.

While this reads well, keep in mind that the language is, like the language in other former Soviet bloc countries, almost identical to the language in the former Soviet Union's speech and press guarantees.

Russia

In theory Russia itself—or, more accurately, the Russian Federation of its many states—established in its 1992 constitution broad-based, cutting-edge democratic freedoms for its media, exceeding those of its former subject countries. This is not surprising, inasmuch as it was Mikhail Gorbachev who initiated the beginning of democratic change for all of eastern Europe in Russia itself, with his policies of glasnost and perestroika. The 1992 constitution stated

Freedom of mass media organs shall be guaranteed. Censorship or monopolies of the media or the abuse of the freedom of mass media organs shall not be permitted. Mass media organs may be established and owned by citizens, public associations, institutions, enterprises, organs of legal self-government, and state organs. Public associations have the right to use state, local, and private radio and television under the conditions and in the procedure established by federal law.

Ironically, early in 1993 Russian president Yeltsin proposed a new constitution for the Russian Federation that included the following article regarding the media:

Everyone has the right to freedom of speech and free dissemination of his thoughts and ideas, and the right to seek, obtain and freely transmit and disseminate information. Freedom of the press is guaranteed. Censorship is banned.

But less than six months later Yeltsin seized control of the media, following his dissolution of the elected parliament, on the grounds that the media had caused the "rebellion" of parliament. When Yeltsin disbanded parliament prior to the rebellion against such an action by many members of parliament, that body was considering telecommunications legislation that would establish a media system and regulatory body similar to that in the United States. Hypocritically, two months later, in December 1993, the new constitution was approved, containing the free speech and press guarantees, even as President Yeltsin maintained full censorship control over the media. The very people in Russia who claim to be the strongest advocates of democracy were, in 1993 and early 1994, the very ones who were stifling democratic freedoms of speech and press. As the new editor of Pravda, Russia's leading newspaper, stated prior to the Yeltsin-ordered new parliamentary elections in December 1993, "We're back to square one now. No parliament, no legal opposition. We have a government-controlled press."

While the principles of speech and press freedom guarantees are stated in most of the new constitutions of countries of eastern Europe, the common factor of a lack of experience with such media freedoms, plus the inability of several countries to practice democracy even as they preach it, has hampered effective implementation.

AFRICA

An essentially third-world continent, Africa is a contrast of strong monarchies, dictatorships, and emerging democracies. But even for the latter, freedom of communication is tenuous, sometimes changing almost daily with the changing of political controls. In a democracy the media serve as watchdogs, informing the public of the bad as well as the good. In many emerging democracies the media serve as propaganda arms of government, under the assumption that such use is necessary to forward the democratic aims of the government. Indeed, many who truly believe in the democratic principle, like Otieno N'donga, a Kenyan serving as director of a communication consortium of five African countries (Kenya, Ethiopia, Tanzania, Uganda, and Zanzibar), points out that a free press and uncontrolled media at this stage in these countries' development could unduly influence the public away from supporting the policies of the respective governments, which are expected to lead to democratic societies. He believes that it is necessary to use the media to unite the people toward common goals, and without the government's use of the media for this purpose, democracy is not likely to be achieved.

Many democratic countries, including the United States, have adopted the practices of totalitarian states from time to time under the guise of fighting totalitarianism. In Africa and elsewhere around the world, some countries that claim to be democracies have even tighter controls over the media than some monarchies and military dictatorships.

Telecommunications in many African states was considered one of the principal tools to help those countries move from colonial or agrarian status to economies that could begin to compete with the west. In the 1960s and 1970s radio and television grew in Africa, but little by little various countries began to see such growth as emulating western approaches and the media were associated in the minds of many with continuing western exploitation. By the 1990s, as noted in Chapter 2, penetration of radio and television was still relatively low: less than 20 percent radio ownership, and less than 5 percent for television. Because in most cases the government controls the media and may change policies and practices at its whim, it is sometimes difficult to find clear operable laws concerning freedom of the speech and press.

Kenya, a leader in the five-nation telecommunications consortium, is one example of the widespread practice of constitutional guarantees that are not necessarily upheld. The Kenyan constitution guarantees absolute freedom to communicate ideas and information. However, such freedoms are up to whoever is president at any given time. The media have learned to practice self-censorship in reporting on the government and its leaders, at the risk of jail or, at the very least, dismissal.

The Congo operates similarly, with its 1979 constitution guaranteeing "freedom of expression, of the press, of association and demonstration." This constitution, as do many others, has a caveat, limiting these freedoms to "conditions laid down by law." The party in power, therefore, decides the extent to which media freedoms are actually applied. Ghana, similarly, has been criticized for claiming that it is moving toward democracy, but in fact, through its ownership of the media, totally controls all media content. Rwanda is another country in this category. Its constitution guarantees freedom of expression, but it is a one-party state with only one radio system, government owned. All journalists must be members of the ruling party, the Mouvement Revolutionnaire National Pour le Developpement (MNRD), and must pledge to contribute to the party's efforts for the growth and development of the nation. Any journalist who does not do so may be sent to jail for up to twelve years. A final example in this category is Zaire. The constitution proclaims "the freedom to print, publish and disseminate written material subject to press laws and relevant regulations in force . . . must not undermine public order, morality and good standards of behavior, or the honour and dignity of individuals." While electronic media are not included in the constitution, from the government's point of view there is no need to put such controls in writing because the government owns and operates all radio and television stations.

South Africa

South Africa could turn out to be a beacon of media freedom as the country moved from minority to majority rule in 1994. The government had tightly controlled all media, allowing no freedoms not specifically approved. During the period of strongest apartheid it took some courageous media to reveal and condemn some of the excesses of government practices. These media and their personnel suffered. In the early 1990s, as the country moved toward ending apartheid, more of the media played key roles in this endeavor. A number of leaders of the opposition parties who became leaders in the new majority-ruled government in 1994 previously spoke out strongly for media freedoms. In the September/October 1993 issue of *Africa Report,* Margaret A. Novicki, in an article entitled "The Press in a Democratic South Africa," quoted African National Congress secretary-general Cyril Ramphosa as stating that "there can be no democracy, at least not in the modern world, without an independent media that is free to inform, to criticize, to probe, and to expose." Novicki also quoted Inkatha Freedom Party president Manosuthu Buthelezi: "the most critical issue, as far as media independence is concerned in South Africa at the present moment, is the matter of broadcasting and in particular of the South African Broadcasting Association. It has been amply proved time and time again what an enormous weapon broadcasting can be in the hands of the government, which is unscrupulous enough to use it as a private propaganda organ."

However, if the new South African government adopts the same policies as other African countries that are ostensibly emerging into democratic states, there may well be no immediate significant changes in freedom for the media. On the other hand, as reported in Chapter 5 by William Siemering, the significant growth of community radio stations is opening the airwaves to access by all groups and opinions.

The critical importance of radio to the political, cultural, and social agendas of people in all countries of the world is illustrated by the reaction in South Africa in early 1994, when the government ordered the closing down of the Radio Pretoria station. The BBC reported that protests took place and that supporters of the station "vowed to defend it with their lives." Whether or not we agree with a given station's political or social purpose, the significance of radio communication is uncontrovertible.

The Mideast

While most Arab countries are geographically located in Africa, they must be considered in the context of the mideast, which includes parts of both Africa and Asia, as established earlier in this book. The mideast, with minor exceptions, universally practices prior restraint, regardless of a given country's constitutional guarantees of freedom of speech and press. We do not suggest that there is any standard of absolute freedom from prior restraint that can serve as a benchmark. While the United States, the British Commonwealth, and the Scandinavian countries come closest to the actual practice of such freedoms, they nevertheless have imposed greater or lesser conditions that in some instances are not too unlike those of some less-than-democratic states.

Prior Restraint

A major difference is that the media in the United States and in many other countries, including the British Commonwealth and Scandinavia, are by and large free to say what they please, subject to specific prohibitions that may be punishable after the fact. In many countries, including those in the mideast, the media are not only subject to specified prior restraints, but perforce are obligated to present favorably the speci-

fied policies of the government or of the religious group controlling the given country.

Prior restraint in the United States relates principally to issues of national security, as enunciated by Justice Oliver Wendell Holmes in his 1919 "clear and present danger" concept in Supreme Court's *Schenk v. U.S.* decision. Withholding of information, which has been done frequently by the U.S. government when damaging revelations might be made about high political figures (such as the stonewalling by President Nixon in the Watergate affair and the refusal to divulge Iran-Contra information by Presidents Reagan and Bush) has the same effect as prior restraint. Nevertheless, prior restraint in the United States is rare compared to its use in most other countries, including those in the mideast.

Censorship

All telecommunications systems in the mideast are censored for the purposes of national security, which includes the protection of government leaders from criticism, and the advancement of religion, which dominates the governments and policies of most mideast countries. For example, Iran's 1979 constitution states that "All civil, penal, financial, economic, administrative, cultural, military, political, and other laws and regulations [including those pertaining to communications] must be based on Islamic criteria." In fact, upon the ascendance to power of the Ayatollah Khomeini in 1979, all radio and television materials were destroyed, and since then the media have been under strict government control to prevent any possibility of their disseminating non-Islamic ideas.

Iraq

Iraq, although an enemy of Iran, is an ally in terms of principles governing the media. The Iraqi constitution of 1968 guarantees freedom of opinion, meeting, and demonstration "within the limits of the law." Media freedoms are nonexistent. A 1968 Press Code forbids any criticism of the president and other top officials. In fact, a journalist who is judged to have insulted the president, a cabinet member, or the ruling Ba'ath party is subject to a penalty of life imprisonment or even death. In 1981 the government promulgated a Ministry of Culture and Information Act, which states that "in the fields of media and information, the Ministry shall propagate and deepen the ideology, principles, and stands of the Ba'ath Party." The government controls all radio and television.

Kuwait

At one time Kuwait had relative freedom of speech and the press, considerably more than most nations in the mideast. Its constitution did include the usual caveat: freedom of the press, but only within "the context of relevant laws." Nonetheless, the print and electronic press had great flexibility. However, in the mid-1980s the ruling royal family disbanded the parliament and suspended parts of the constitution after a confrontation somewhat similar to the 1993 Yeltsin-Russian parliament disagreement. Following this event a system of prior restraint was put into effect for all media, and the minister of information was given complete control over both electronic and print communications.

During and following the Gulf War, control of speech and media grew tighter, in some respects even exceeding Iraq's restrictions of these freedoms, and prompting some observers to question the validity of the U.S. and UN Gulf War efforts on the grounds of achieving freedom for Kuwait.

Egypt

Egypt is considered one of the more moderate countries in the mideast; yet some laws restricting the media are as harsh as those in dictatorial and monarchial nations. The constitution appears to forbid censorship and guarantee fredom of speech and the press. But the government reserves the right of prior restraint and/or postdissemination punishment for airing "anything prejudicial to national unity and social peace . . . scurrilous material that might offend the state and its constitution . . . to challenge the truth of divine teachings . . . to advocate opposition or hatred of state institutions . . . to publish abroad false or misleading news or information which could damage the interests of Egypt." The radio and television systems are government run and the president of Egypt has absolute power over all media. While upholding the concept of freedom of the press, the Charter of Honour of the Radio and Television Union also proclaims the necessity for media practitioners to strengthen the principles of nationalism and to promote religious freedoms, among other things. Violators can be suspended from their jobs.

Keep in mind that Egypt, along with other countries that desire to achieve democratic governance, looks upon its control of the media as a positive requirement for the good of its people. Some defenders of such practices in these countries point to the fact that the United States practiced total censorship and prior restraint of American journalists covering the Gulf War under the same pretexts.

Jordan and Lebanon

Jordan and Lebanon are among the most liberal of the mideast Arab countries in terms of media control. Although the Jordanian government controls radio and television, action may not be taken against media personnel without a due-process court trial. Lebanon, as discussed earlier, has an extensive system of privately owned media. However, individual journalists have been arrested and stations have been closed down for disseminating material considered by the government to be not in the public interest.

Israel

While Jordan is quickly moving toward western principles and has an elected parliament—though under the ultimate jurisdiction of the king—it is Israel that is usually thought of as having the most western-oriented telecommunications system. The two Israeli television stations are headed by government-appointed Director-Generals, and although there is no direct official systematic censorship or prior restraint, the television and radio stations practice their own, especially in the critical area of national security. Media practitioners are enjoined to promote the welfare of the country and its education. They are warned not to carry opinionated commentary and to avoid "infotainment" news coverage such as glamorizing demonstrations and other political happenings. Inasmuch as the media are government operated or supervised, such variations from media freedoms are generally adhered to.

Ostensibly to avoid fanning the flames of controversy, key issues and occurrences are frequently ignored or played down. Although specific incidents of Israeli harsh action against Palestinians are reported, the electronic media have not attempted to discuss the country's apartheid practices in general or given much coverage, except during the new Labor government peace process efforts in the early 1990s, to the majority of Israelis who have supported peace and accommodation with the Palestinians for decades. The courts have on a number of occasions denied the government's attempts to censor media material, although such restrictions on censorship did not apply to the Israeli military's strict control of Palestinian media in the West Bank and the Gaza Strip (or, in Arab terms, the Occupied Territories). A media double standard existed, and continues to exist, in the areas that were not granted self-rule in the Israel-PLO agreement of 1994.

ASIA

Japan

For years Japan has been the leading telecommunications power in Asia, and in fact, the leading distributor of telecommunications equipment in the world. However, as China moves ever faster toward an international economic and market role in the 1990s, so too is its telecommunications system growing. The two former powers of post–World War II Asia appear once again to be becoming mutually strong, this time in respect to the entire world. The international growth of many aspects of human endeavor is reflected in and affected by the international growth and interchange of telecommunications. Both China and Japan are examples of this trend. Yet so close geographically, they are worlds apart in terms of media freedom and control.

Japan exercises little direct control over the content of media, even through its funding of NHK, the Japanese national broadcasting system, although its constitution provides for a general exception: "the freedom and rights guaranteed to the people by this constitution shall be maintained by constant endeavor of the people to refrain from any abuse of these freedoms and rights and shall always be responsible for utilizing them in the public welfare." While this opens up to interpretation the guarantees of freedom, Japan has by and large stood by the basic principle of its declaration.

Before and during World War II Japan exercised tight censorship of the media. Its Broadcast Law of 1950, developed under U.S. occupation, reflects the U.S. Constitution in its guarantee of freedom of speech, and adds such areas as impartiality and tolerance. Note that the broadcast law applies principally to the government system, NHK, and that commercial broadcasters are simply strongly encouraged to comply. As in the United States, principal inroads on the basic freedoms of the media have been in areas of indecency and slander. Japan does practice one aspect of control usually found in nondemocratic states: membership in a journalist association is required for accreditation and access to certain government information; these associations have self-censorship restrictions that reporters agree to abide by. Self-censorship also reflects government standards in a manner similar to the U.S. ban on cigarette commercials and the oversight by the Federal Trade Commission of false or misleading commercials. Advertised medicines and cosmetics, for example, must be in compliance with the standards established by the Ministry of Health and Welfare.

CHINA

In China, as noted earlier, all telecommunications is under the direct control of the government, and the myriad of television and radio stations throughout the country, although distant and frequently geographically independent of Beijing, are expected to follow the policies set down by the Central China Television (CCTV) and China People's Broadcasting Station (CPBS), both under the control of the Ministry of Radio and Television. Policies for the media are determined by the Central Committee of the Communist Party. While all media content is ostensibly supposed to forward the policies of the government, in the early 1990s, concomitant with its changing economy and growing international role, China began to diversify media content, which included the importation of some foreign programs through satellite, and greater flexibility for individual regional and local stations in developing their own programming. However, the basic government controls still remain, and criticism of the government leaders or their policies is still not found in Chinese media. In some cases even the foreign media are silenced. For example, in 1994 hotels in Beijing were ordered to turn off distribution to hotel rooms of the satellite-received CNN international news network from June 2 until June 6. The government was concerned that CNN would carry file footage of the suppression of the Tiananmen Square democracy movement on its fifth anniversary, June 3 and 4.

Hong Kong under British control (until 1997) does not have the freedoms accorded media in the United Kingdom. Hong Kong does not even have a constitutional guarantee of freedom of the press. The government may exercise prior restraint and censor all programming. In fact, much of the news broadcast on Hong Kong media comes directly from the Hong Kong Government Information Service (GIS). Many Hong Kong journalists have bought into the rationale that it is in the public interest to concentrate on the "good" news and to omit or play down any "bad" news that might embarrass the government.

South Korea

After years of military rule, South Korea's emergence into a fledgling democracy placed it in a position similar to that of many politically changing countries. While it proclaims freedom of speech and the press, the government maintains strong influence over content on the grounds of strengthening the nation's unity and stability. All stations, in conjunction with a government-established committee, must review program content both before and after transmission. The government's Bureau of Broadcast Management establishes criteria for programming and makes sure that journalists promote national loyalty and pride. In one important respect it provides a more democratic base than do many countries, including the United States: it mandates equal time for all candidates for federal office, regardless of a candidate's wealth and ability to buy air time.

India

Although the constitution of India does not have a provision for freedom of the press, India's courts have frequently acted to ensure such freedoms under the broader rubric of the constitution's freedom of expression. However, the volatility and rebellious factions that have marked India's democracy since its independence have prompted the courts, as well, to sometimes subordinate freedom of expression to its perception of the need for public order and the curbing of public unrest. Further, a carryover from British colonial rule is an Official Secrets Act that limits journalists from investigating matters relating to the government. When India developed electronic media it placed all telecommunications under the government's Ministry for Information and Broadcasting. The government even censors videotapes made available for public rental. Because of the government control of the electronic media, journalists, with rare exceptions, present the government view of matters and usually ignore opposition concerns. The basic code for broadcasting is the Indian Wireless Telegraphy Act of 1933, which was developed under British rule. This act has specific prohibitions, such as attacks against the state, religions, a political party, individual political figures, and the airing of what might be deemed indecent or libelous material. Final arbitration of violations lies with the Ministry of Information and Broadcasting, placing close control over content in the hands of the government.

Indonesia

Indonesia is representative of those countries, especially in the Pacific area, with geographically scattered populations. Indonesia's thousands of islands can be reached effectively only through radio and television. Rebellious factions in Indonesia are used by the government as justification for its domination of both the television and radio industries. The Ministry of Information not only provides acceptable news for the media, but has been known to contact journalists working on stories and strongly suggest ways in which the stories might most effectively be covered. As in many other countries, the constitution states that "freedom of speech and the press . . . shall be provided for by law." And, also as in many other countries, there are conditions such as a requirement that the press "fan the spirit of dedication to the nation's struggle . . . strengthen national unity and integrity . . . exercise social control where it is constructive."

The countries above are more or less illustrative of the degree to which media freedoms are operative in Asia. By and large, the media are under strict government control or, at the least, strong government supervision or influence. Perhaps this control, the result in some cases of the perceived need to strengthen the unity and stability of an emerging democratic state, and the desire to establish efficient and peaceful relationships with neighboring countries, will abate into the granting of more freedoms to the media as each nation's goals come closer to achievement and working democracy can actually be practiced on a larger scale.

OCEANIA

While Australia's and New Zealand's broadcasting systems are similar to those of the United Kingdom and other British Commonwealth countries, providing government support but a minimum of government control, some of the smaller island nations of the Pacific basin exercise strong government control. The principal reason is that the multiple far-flung islands that make up some of these countries—in some cases hundreds and even thousands of separate islands—requires centralization of signal distribution in order to reach all of the population. In some instances the geographic rationale is coupled with the instability of the government and its need to use the media to solidify its standing among the people.

Cultural taboos and local news play important roles in determining the programming in most of the island nations. For example, PACBROAD, the satellite news and program exchange service for the Pacific Islands, is carefully monitored and programs that are not consistent with the social or political policies of the receiving nation are excluded. Family planning, for example, is taboo, and few islands receive such programming, even from international sources such as the UN. Further, the small populations and, in many cases, limited geographic areas of a given island nation result in stressing news approximating that of the weekly newspaper in small-town America: weddings, funerals, and travel, with little international or even regional news.

Australia

The Australian Broadcasting Corporation is funded by Parliament, but its Board of Directors ostensibly operates free of government interference to the same extent that the BBC does. Each station develops its own policies. However, all broadcasting is responsible to the Australia Broadcasting Tribunal (ABT), which functions as a regulatory body. It establishes standards and if any of its proposed rules impinges on program content, it holds public hearings on the matter. The ABT does have the authority to censor programs, including foreign imports, it considers inimicable to Australian morals or culture. Counter to this, a code put forth by the Australian journalists' association promotes objectivity in content and the public's right to know. However, there are no enforcement powers in this voluntary code. Parenthetically, Australia's media, which serve many of the nations in the Pacific area through satellite (as described in Chapter 2), have been criticized by a number of the region's nations, such as Papua New Guinea, Sing-

apore, and Fiji, for what they believe is "colonial" and "racist" material. Some nations have called for regional program content guidelines.

New Zealand

The Broadcasting Council in New Zealand functioned for many years somewhat similarly to the ABT, and the broadcasting system also resembles that of the BBC. However, in the late 1980s the government began a policy of privatization, in which many of the country's public utilities, including broadcast frequencies, were sold to private parties, many of them foreign investors. While the government maintains supervision over its own systems, Radio New Zealand and Television New Zealand, its regulatory hold, especially over private stations, has decreased.

Fiji

Fifi's broadcasting was originally set up by its colonial ruler, Great Britain. As do other former territories in the Pacific, Fiji continues to maintain government supervision over content. In part, government control is a legacy of the time Britain occupied these countries and used the media to forge its governing policies. One rationale given for government control over content is that it is one means of creating a system and viewpoints independent of British and neighboring Australian influence. Another factor in government supervision of program content in many of these countries is their status as developing nations, prompting the parties in power to include the media as a tool in moving up from third-world status.

Xenophobia

Fear of foreign influence or outsiders affects all countries in their judgment of the freedoms appropriate for, and the responsibilities of, the media. Most countries do not hesitate to criticize other nations while either finding rationale to laud, or altogether ignoring, their own shortcomings. For example, on a 1993 "Nightline" program the American Broadcasting Company correspondent in Beijing appropriately stated that Chinese television was being used to aid the government (for example, sometimes publicly identifying dissenters) rather than what it should be used for, to enlighten the people. Few who believe in democratic freedoms of the media can argue with this criticism. However, it is indicative of xenophobic vision that the same correspondent did not deem it appropriate to note that the media in ostensibly democratic nations frequently are used by their governments in the very same way.

7 External Services and Organizations

EXTERNAL SERVICES: RADIO

Almost all first-world countries and even many third-world countries have some form of external broadcasting service. These range from global coverage to coverage of limited areas of the country's geographical region. Several factors account for a nation's allocation of funds for capital expenditures and operations of an external service: political propaganda, cultural expansion, trade and investment promotion, and promotion of tourism. The political purpose usually is primary: to create or enhance an image of the given country to other countries.

In nations controlled by royal families or military dictatorships, the external service, like the domestic service, is used as a public relations device, to build up the support or at least acceptance of those in power in order to facilitate their maintenance of such power. In democratic countries the external service is principally a foreign relations tool and, usually, is under the control of the Ministry of Foreign Affairs or, as in the United States, the Department of State.

With the end of the almost fifty-year cold war in the 1990s, a number of countries have had to take a new look at their external services. The United States, for example, was in the process of phasing out its taxpayer-supported Radio Free Europe (RFE) and Radio Liberty (RL) as this is being written. RFE and RL were aimed at the Soviet Union and its eastern European allies, so there was no further political justification for these stations. In addition, those eastern Europeans who had previously listened to these stations as a supplement or alternative to the news they received on their domestic stations now had, in many cases, their own stations free of government control, and they no longer had an interest in or need for the U.S. broadcasts. In the mid-1990s RFE and RL were still operating, but at reduced levels. Further, Radio Marti, a controversial U.S. external propaganda service directed to Cuba that has been questioned since its inception, was also coming under serious new review as this is being written. In mid-1995, however, Radio Marti was also still in operation.

External services are principally government-funded and -operated shortwave radio stations, but include religious stations and even some private commercial stations. In recent years satellite technology has made possible the development of some external television services, but in the mid-1990s radio was still the overwhelming external service medium. Many countries have multiple services, some with many stations reaching the same target area. Most broadcast in several languages—some as many as forty or fifty languages—aimed at continental areas and individual countries where the given language is spoken. For example, a country may have a service covering most of Asia and additional separate services directed specifically at South Asia, Southeast Asia, Southwest Asia, and East Asia; or to North America, Latin America, Central America, and South America; or to North Africa, Northwest Africa, Central Africa, West Africa, East Africa, South Africa, Southwest Africa, and the Middle East, as well as Africa as a whole. Some services target individual countries. Cuba, for example, has an external service aimed at eastern North America. A number of countries, continuing their cold war propaganda as peacetime propaganda, have external broadcasts aimed at Russia, the Ukraine, and other former Soviet Union or Soviet bloc nations.

Broadcasts in a given language are not necessarily directed only to places where that language is the native tongue. English, which in recent decades has replaced French as the international language, is the most widely spoken second language in the world, and there are hundreds of English-language external services from many countries, aimed at every continent and every subregion in the world.

Countries with external services need multiple channels for the same station or service because of the limited number of channels globally. Channels are assigned by the International Telecommunications Union, as described elsewhere in this book, and nations frequently have to share time on the same channel. A given service—let's say English-language broadcasts from Radio Nederland on the Netherlands external service directed to South Asia—may use one frequency for several hours, shift to another in a few hours, and later shift to a third, or even additional frequencies.

The economy plays an important role in external services operations. As already noted, industrialized first-world nations that can afford such services have them, some in great degree, others with minimal facilities. Developing nations, as shall be seen in an overview of continents and countries later in this chapter, rarely can afford to have them.

The worldwide faltering economy of the early 1990s impacts rich and poor countries both, but in different degrees. In 1993, for example, Estonia's economy forced it to suspend its external service. Budgets for the United States' Voice of America (and Radio Free Europe and Radio Liberty), the BBC's World Service, Germany's Deutsche Welle, Radio Moscow, and Spanish Foreign Radio—five of the world's leading services—were cut back. Countries with more limited services, such as Radio Prague, were also cut back. Radio Vilnius' extensive service was not eliminated, as was that of its Baltic neighbor, Estonia, but was curtailed. One result of the global economic problem was that a number of countries began to cooperate in their use of facilities in order to make more efficient use of their external service budgets—for example, BBC and Japan, Radio France and Swiss Radio International, and Radio Nederland and Belgium's Radio Vlaanderen. Probably the most respected and wide-reaching religious organization global service, the Christian Science Monitor World Service, not only suffered budget cuts, but in an attempt to find a broader base of support changed its name to Monitor Radio International.

But even as most countries pulled back, there was expansion in some places. A new shortwave external service, Australian Armed Forces Radio, went on the air in 1993, and the United States' Voice of America expanded its coverage by adding two new transmitters, in Thailand and Morocco.

The British Broadcasting Company's World Service (BBCWS) is generally considered better than that of any other country in its quality, depth, coverage, and objectivity. It provides service to every continent in the world in virtually every significant language, as well as a comprehensive and globally extensive English-language service. The United States' Voice of America, Radio Moscow, and Radio Beijing are considered to be the other principal external services in the world. In the second tier would be

Germany's Deutsche Welle, the Netherlands' Radio Nederland Wereldomroep, Radio Canada, Radio France International, and Radio Exterior España.

A private commercial external service, CNN, was the first worldwide news service. In 1994 BBC's World Service was expanding so as also to provide global coverage with its news. The following examples from the various regions of the world indicate the scope of shortwave radio external services.

North America

The United States' Voice of America (VOA) operates over sixty transmitters from the contiguous states, plus relay stations all over the world, including Albania, the Ascension Islands, Belize, Botswana, Bulgaria, the Czech Republic, Germany, Greece, Kuwait, Morocco, the Philippines, Portugal, Romania, Russia, Sao Tome, Sri Lanka, Thailand, United Arab Emirates, and the Uunited Kingdom VOA covers the entire world in more than 45 languages. In addition to the previously mentioned cold war adjuncts, Radio Free Europe, Radio Liberty, and Radio Marti, the United States has a continuing operation by its Armed Forces Radio and Television Service (AFRTS). Its broadcasts are principally on open frequencies (except in the United States) and can be heard and sometimes seen in the immediate and adjacent countries from which they operate. The Armed Forces Network's (AFN) shortwave radio broadcasts can be heard internationally. One AFN service was a casualty of peace. Since the end of World War II, the AFN radio station in Berlin served the people of West Germany and many in East Germany who listened surreptitiously. With the final withdrawal of American troops from Berlin in 1994, AFN Berlin was discontinued. At their peak, RFE and RL broadcasts were in twenty-three languages to east, central and southeast Europe, the Baltic states, Afghanistan, and the former Soviet Union countries

(later members of the Commonwealth of Independent States, or CIS).

In addition to these government-operated external services, a number of religious radio external services broadcast from the United States, including Catholic, Protestant, Latter-day Saints (Mormon), Fundamentalist, Christian Science (Monitor Radio), and the Assemblies of Yahweh. Alaska is the site for one of the larger religious external services. The New Life Station of the World Christian Broadcasting company aims its broadcasts at Russia and eastern Asia. Most of the external services operated by religious groups are missionary and proselytizing in nature. In addition, several private commercial radio external services exist. The United Nations has its own external radio service that can be heard throughout the globe from its headquarters in New York. The UN leases facilities from VOA, with some of its programs aired via satellite, but most through tape distribution.

Radio Canada International uses over fifty frequencies and concentrates its broadcasts to Europe, Africa, Asia, the Middle East, Latin America, the United States, the Caribbean, and to two eastern Europe countries, Russia and the Ukraine.

Cuba is the only Central American country with meaningful shortwave external broadcasts. Radio Habana Cuba concentrates its broadcasts to North, Central, and South America, Europe, Africa, and the Middle East.

South America

Only a few countries in South America have external services, mainly because most simply cannot afford them. Argentina's Radiodifusion Argentina Al Exterior has only three transmitters and broadcasts principally to North and South America, Africa, Asia, and the Middle East. Brazil also has only three transmitters, with its programming aimed at North and South America, Europe and Africa. Ecuador's La Voz De

Los Andes is somewhat larger, with thirty transmitters carrying programs worldwide, including eastern Europe and former Soviet Union countries now part of the Commonwealth of Independent States (CIS).

Europe

Europe has the largest proliferation of external services of any continent. Almost every country has some kind of external broadcasting system. As already noted, the BBC World Service in English and most every other language reaches virtually every corner of the globe. In western Europe, Belgium's Radio Vlaanderen International serves North and South America, Africa, Asia, Europe and the Middle East. Radio France Internationale covers every continent, including the Pacific Basin, and has special services for subregions of individual continents. It also aims some services at individual countries, such as China, and French-speaking countries such as Laos. Germany's Deutsche Welle also reaches every continent, including subareas. It has a large number of services aimed at individual countries, including nations in eastern and western Europe and Asia. Radio Nederland also covers every continent in the world, with a number of services aimed at subcontinent areas. It provides special programming to Australia and New Zealand. Swiss Radio international doesn't have to target European countries because a number of its domestic foreign language stations' signals reach other nations in Europe. Its external service, however, reaches every other region of the world.

The Scandinavian countries' external services are not as extensive as those of some of their neighbors to the south. Radio Finland covers North and South America, Africa, east Asia, the Middle East, northern Europe, and Australia. Radio Sweden broadcasts to some part of every continent. Radio Norway International does the same. But their schedules are limited. Radio Denmark uses Norway's facilities to broadcast its programs, in Danish, to its nationals living in other lands.

In southern Europe Spain has the leading external service. Radio Exterior de España broadcasts extensively to every continent except Asia. Radio Austria International broadcasts to every continent and the Pacific Basin, but on a more limited schedule. Radio Portugal's service is directed to every continent except Asia, with special services to countries with substantial Portuguese-speaking populations: Brazil, Venezuela, and the United States. The Voice of Greece concentrates on various regions of Europe and Africa, on North, Central, and South America, and on the Middle East. It also has special programming aimed at individual countries, including neighboring nations Cyprus and Turkey, Japan in the Far East, and CIS countries. The Voice of Turkey broadcasts to North America, Asia, Africa, Europe, and the Middle East. Italy's RAI broadcasts to North America, Latin America, Africa, the Near East and Far East, the Mediterranean, and has a special European service. Malta operates the Voice of the Mediterranean. The Vatican operates its own external service, broadcasting to Europe, Africa, the Middle East, the Pacific Basin, Asia, and North and South America in various languages.

In eastern Europe Russia has long had the dominant external service. Radio Moscow International broadcasts to every continent in the world in every principal language and in many lesser-used languages, similar to the operation of the BBC World Service. It also has services aimed at subcontinent regions. Radio Moscow is still heard extensively throughout the United States, having been a key propaganda organ for the USSR during the cold war just as VOA was for the United States. In addition to Radio Moscow, Russia operates the Voice of Russia, a world service in the Russian language.

Albania's Radio Tirana has surprisingly large coverage for a small, poor country, reaching Europe, the Middle

East, North, Central, and South America, and Africa. Bulgaria's National Radio Foreign Service is directed at all areas except Asia and the Pacific Basin, and has some services aimed at individual countries. The Czech Republic's Radio Prague reaches part of every area of the world. Slovak Radio International directs its external services to North and South America, Europe, and the Pacific. Poland's Polskie Radio broadcasts principally to European countries. Radio Romania International not only reaches every continent, but directs special programming to such individual states as Russia, the Ukraine, Australia, New Zealand and Hong Kong. Hungary's Radio Budapest aims at Europe, North America, Asia, and the Middle East. Even during the bloody conflict in former Yugoslavia, Radio Yugoslavia in Belgrade broadcast its external services to Europe, Africa, the Middle East, North America, and the Pacific.

Radio Ukraine directs its external services to Europe and North America, and uses Radio Moscow's relays for broader coverage. Belarus's Radio Minsk broadcasts to North America and Europe. The target area of Radio Moldavia International includes South America as well as North America and Europe. Lithuania's Radio Vilnius reaches Europe and North America. As noted earlier, Radio Estonia suspended operations in 1993.

Africa

A number of even the poorer African countries have some external services that broadcast to other African nations. Ethiopia has a limited service, including English-language programs. So does Ghana. Libya's limited service broadcasts principally to Eastern Europe. With Malta, Morocco operates Radio Mediterraneé Internationale. Mozambique has a limited external service in English. The Voice of Nigeria principally serves West Africa, and includes English- and French-language programs. The Sudan similarly has a limited service, including English and French broadcasts. Tanza-

nia's limited external service is directed toward eastern, central, and southern Africa. South Africa aims at other African nations with its Channel Africa.

Several countries do not have governmental external services, but serve as bases for shortwave stations of religious organizations. For example, one of the larger groups, Far East Broadcasting Association, sends its religious programs to Asia, Africa, and the Middle East from the Seychelles. Another large religious organization, Trans World Radio, broadcasts to African nations from Swaziland. A number of African countries have relay stations for the larger external services of other countries such as Germany, France, and the Netherlands. Ascension, for example, is used for relay stations by both BBC and VOA.

Many mideast countries have shortwave external stations. Egypt has relatively extensive broadcasting, reaching every continent and the Pacific Basin. The external broadcast service of the Kingdom of Saudi Arabia reaches Africa and Southeast Asia. The external broadcast services of the Syrian Arab Republic send their signals to North and South America, Europe, the Pacific Basin, as well as the Middle East. The Islamic Republic of Iran Broadcasting directs its programming to Asia, Africa, Europe, and North America, in addition to the Middle East. Radio Iraq International programs to North America, Africa, Europe, and also to the mideast region. Israel does not do special programming for its external service, The Voice of Israel, but relays selected domestic radio programming to North America, Europe, Asia, Africa, as well as to other Middle East nations. It includes an English language service.

Asia

China's external service is one of the most extensive in the world. China Radio International broadcasts globally in more than forty languages. Radio Japan uses more than sixty different frequencies to reach every continent and region in the world. All India Radio, while more limited, nevertheless reaches out to all areas except North and South America. North Korea's Radio Pyongyang broadcasts to North and Latin America, Europe, Africa, and the Middle East, as well as to Asia. South Korea's Radio Korea reaches North America, Europe, Africa, and the Middle East, in addition to Asia. Radio Pakistan programs to Europe, Asia, Africa, and the Middle East. Mongolia's Ulan Bator Radio broadcasts to Europe, Asia, the Pacific Basin, and the CIS. Sri Lanka sends its signals to Asia, North America, Europe, the Pacific Basin, and the Middle East. Taiwan operates the Voice of Free China and the Voice of Asia. The former broadcasts to North, Central and South America, Europe, the Middle East, and to Asia. The Voice of Indonesia, though limited, broadcasts in ten languages, including English. Bangladesh and Cambodia, extremely poor, nevertheless operate small external services, and include English language programming. Laos has a very limited service, but still offers programs in five languages. Radio Afghanistan broadcasts to Europe, Asia, and the Middle East. The Voice of Vietnam sends its signals to North and Latin America, Europe, Africa, and Asia.

Countries that were formerly part of the Soviet Union and, in some cases, are now members of the CIS, are sometimes included in discussions of eastern Europe as an extension of their former control by Russia. However, these countries are now, in fact, independent nations within Asia. Armenia's Radio Yerevan joins with the Araks Radio Agency to send programs to North and Latin America, Europe, and the Middle East. Azerbaijan's Radio Baku broadcasts to the Middle East. Radio Georgia aims its signals at Europe and the Middle East. Uzbekistan's Radio Tashkent limits its external broadcasts to Asia and the Middle East, and Turkmenistan to the Middle East only. Kazakhstan's Radio Alma Ata principally serves Asia.

Oceania

The external services of nations in this area are quite limited. Australia aims only at Asia and the Pacific with its external service, Radio Australia. Like many other countries whose radio signature theme, such as the BBC's music and tone logo, can be recognized instantly, Radio Australia starts its broadcast day on all of its frequencies with five minutes of the music internationally associated with the country, "Waltzing Matilda." Radio New Zealand International has only a few frequencies and limited hours for its external service, broadcasting mainly to the Pacific islands. The Philippines has Radio Filipinas, Radio Veritas, and the religious Far East Broadcasting Company sending their signals principally to countries in Asia and the Pacific.

The U.S. presence adds to the external service mix in the Pacific. Hawaii is home to the religious external broadcasts of World Harvest Radio, reaching Pacific Basin countries and eastern Asia. The U.S. territory of Guam accommodates two religious shortwave broadcast systems directed to the Pacific island nations and Asia, Adventist World Radio and Trans World Radio Pacific. The Northern Mariana Islands, a U.S. commonwealth, has the Far East Broadcasting Company sending religious programs to the Pacific islands and Asia, and Monitor Radio International broadcasting to Asia, Africa, the Pacific, the Middle East, and (in eastern European languages) to eastern Europe.

EXTERNAL SERVICES: TELEVISION

Television external services were still essentially in the developmental stage in the mid-1990s. Radio was still by far the principal external service broadcast medium. Nevertheless, the outreach of BBC World Service Television and the success of a private global news service, CNN, plus the continuing rapid development in satellite-transmission and -reception technology suggested that by the beginning of the 21st century television would be well on its way to becoming an important part of many countries' external services. In North and South America external broadcasting by some private companies was on the upswing in the early 1990s, although their purposes were not related to the traditional external service goals, but to expanding their commercial bases. Cooperative organizations and groups have experimented with external video services. For example, the Union Latinoamericana y Caribena de Radiodifusion (ULCRA) was in the early 1990s producing a weekly TV news review, "El Latinoamericana," for distribution to about a dozen countries. Some regional broadcasting associations were doing joint productions for international sharing.

In Europe, as discussed earlier, the closeness of countries automatically results in the international distribution of television programs, though through domestic channels. Radio-Television Luxembourg is a prime example. Some channels were deliberately oriented to external regions, such as Scanset and Nordisk to Scandinavia, 3-Sat to German-speaking countries in Europe, and La Sept to French-speaking areas. As noted earlier, BBC World Service TV covers all of Europe as well as some other parts of the world, as does Rupert Murdoch's SKY Channel, and both are moving toward global coverage.

In Africa and the Middle East there was no significant external TV programming on the immediate horizon in the early 1990s, except for the already mentioned directing of specific programs to neighboring countries on domestic TV channels whose signals cross national borders. In Asia and the Pacific the principal TV external service is through the Pan-Asian satellite network, utilizing four channels, with STAR TV from Hong Kong carrying BBC World Service TV programming. The Asia-Pacific Broadcasting Union was encouraging and facilitating TV program exchange in the

early 1990s, but as this is written there was not yet a formal regional external service.

ORGANIZATIONS

As telecommunications have advanced, we have come closer and closer to the global village that was predicted so confidently at the advent of satellites and other new technologies in the 1960s. International communications and the exchange and co-production of programming on regional and world levels have spurred the growth of organizations designed to facilitate the global telecommunications village.

Policy Groups

ITU

On the international level the oldest and most comprehensive organization is the International Telecommunications Union (ITU) (Figure 7-1). The global organization that eventually became the ITU was established in 1875 as the International Telegraph Union to deal with the expanding international use of the telegraph and, a few years later, the telephone. Its beginnings go back as far as 1849, when German states and Austria signed the first of the world's communication treaties, regarding the recently developed telegraph. Through several metamorphoses, including the development of wireless communications, worldwide expansion of ship-to-shore communications, and the founding of a Radiotelegraph Union, the International Telecommunications Union came into being in 1932. Its jurisdiction included telegraph, telephone, and radio. Consisting of 170 countries as of the early 1990s, the ITU holds Plenipotentiary (its highest level) meetings every few years to review problems and new technologies. Its principal function is to do on a worldwide scale what the Federal Radio Commission was set up in 1927 to do in the United States, and what many similar

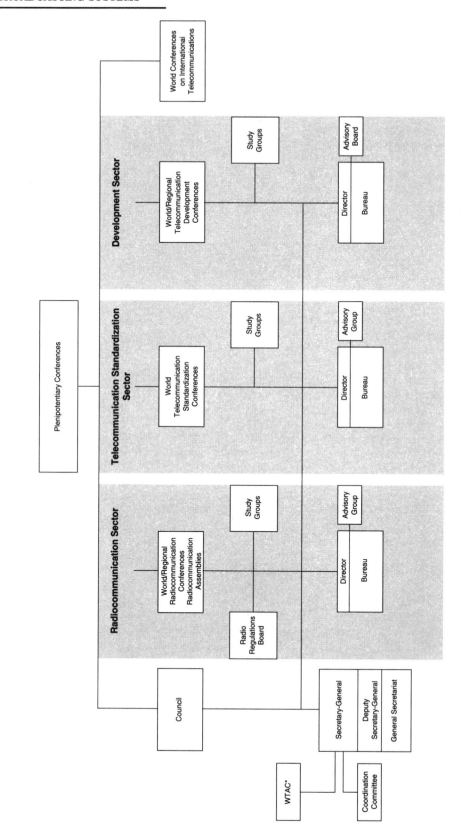

FIGURE 7-1
ITU organizational
chart Courtesy ITU.

regulatory bodies were set up to do in other countries: allocate frequencies on an equitable basis in a manner that eliminates any possibility of signal interference among any countries. As a specialized agency of the United Nations since 1947, the ITU's philosophical purposes are to apply to telecommunications the same principles the UN applies to other endeavors: to promote peace and the economic and social development of member countries. In this respect, it attempts to assist developing countries in their development of telecommunications facilities and operations. Its stabilizing efforts include the setting of international technical standards, universal definitions of telecommunications terms, including those for new and emerging technologies such as computer data exchange, and the development of guidelines for the lowest possible tariffs among nations. It has two major divisions: the International Frequency Registration Board (IFRB) deals with technical standards and frequency assignments and use; the International Consultive Committees (ICCs) work with policies and rules, including enforcement procedures.

A key arm of the ITU, which schedules working conferences every ten years, but holds periodic special meetings that make its work of an ongoing nature, is the World Administrative Radio Conference (WARC). The term *radio* as used in the legal and international context is a generic one, referring to all over-the-air communications. WARC's principal function is to assign frequencies to regions and individual countries. With the development of satellite communication, its work became truly international. It has to deal with cross-border distribution that goes far beyond contiguous countries and even regions by designating geostationary orbits for communication satellites. For many years the countries with advanced technologies and resources for satellite communications, such as the United States, the Soviet Union, Japan, and some nations of Europe, used the limited number of frequencies allocated for satellite use because they were ready to go. Third-world and other developing countries were not able to get a "bird" into the air. However, as these developing countries began to join together to strengthen their international political power, they also began to pressure and in some cases outvote the larger nations that had controlled WARC policies, and during the past two decades have claimed a share of frequency space, if not for immediate use, as reserved frequencies for future use. At the 1980 WARC conference, for example, many of the Latin American countries banded together to restrict the growing U.S. monopoly of western hemisphere frequency allocations.

ITU's operational structure is headed by its Plenipotentiary Conference, at which its member nations meet every seven years or so. The conference reviews the International Telecommunications Convention under which it operates to determine whether any changes need to be made in policy or implementation. It also selects the ITU Secretary-General, Deputy Secretary-General, the five members of the International Frequency Registration Board (IFRB), and two directors for the International Consultative Committee. The day-to-day management of the ITU is under the Administrative Council. The Council is responsible for the periodic meetings representing the two major areas of its work: the Administrative Telegraph and Telephone Conference, and the Administrative Radio Conference. The General Secretariat, headed by the Secretary-General, coordinates the work of the International Frequency Registration Board (IFRB), the International Radio Consultative Committee (CCIR), and the International Telegraph and Telephone Consultative Committee. Funding for the ITU comes in part from contributions of its membership and in part from the UN, principally the United National Development Program (UNDP), which provides support for ITU assistance to developing countries.

At the end of 1992 the ITU instituted some important changes in its operations, reflecting the changing political, social, and economic conditions of the world. Traditionally, the Plenipotentiary conferences were limited to direct participation by official government representatives only. Now the debates and forums of the conferences and the frequent sometimes yearly subconferences are open to nongovernmental representatives, particularly from the business, industry, and corporate fields. The ITU recognizes the rapidly expanding growth of communications technology and the increasing critical role of the private sector in the imminent global information superhighway.

Another structural change related to mission. At one time the ITU's Technical Bureau for Development's role was more passive than active. Today it is an integral part of the Secretariat and the ITU's budget, providing increasing technical assistance to member countries.

Further, the ITU altered its traditional attempts to establish worldwide technical standards—which its standard-setting bodies were increasingly unable to do because of the rapid growth of new technologies. It now attempts to coordinate the various standards throughout the world.

By 1994 the ITU had also established a global information infrastructure, using Internet to make information on its activities available to anyone worldwide.

UNESCO

The United Nations Educational, Scientific and Cultural Organization (UNESCO) is associated in most people's minds with human rights and the preservation and growth of culture. Recognizing the role of communications in achieving its major purposes, UNESCO has become a major international supporter and adviser for telecommunications growth in many countries of the world. It provides consultants and training, arranges for facilities, and coordinates program exchange, mainly for third-world countries. To concentrate on developing nations, UNESCO established the International Program for the Development of Communication (IPDC).

World Bank

The World Bank provides funds to developing countries to build up infrastructures in key areas of transportation, energy, and communications. However, the operation of the World Bank has traditionally fallen to conservative managers who rarely recognize the innovative areas that do not conform to their traditional thinking. Consequently, telecommunications, although known by most of the leaders of the world to be the key elements in a country's political, cultural, and educational development, have received relatively little support from the World Bank.

GATT

While the international General Agreement on Tariffs and Trade (GATT) of 1993 is not an organization, it nevertheless has a significant impact on global telecommunications. While establishing free trade among countries, it exempted television programs, allowing individual countries to set import quotas. As the largest distributor of TV programs, the United States strongly dissented from this exemption, but in order to see the overall pact approved, eventually agreed to its inclusion. Principal supporters of GATT were the European nations who, through the European Commission (now the European Union, or EU), established the European Directive described earlier, limiting the percentage of foreign TV products that could be imported by EU members.

Satellite Groups

INTELSAT

Other than UN-coordinated organizations that deal with law, policy, and international development, the major global organizations are those that relate to the principal means of international communications distribution, satellites, and those that relate to program exchange, which are discussed later in this chapter. The principal satellite organization is the International Telecommunications Satellite Organization, INTELSAT (Figure 7-2). In 1994 it had a membership of 119 nations from all parts of the world. It was established in 1964, in time for the launch of the world's first communications satellite, "Early Bird," in 1965.

INTELSAT operates the space aspect of the system for its members, including the satellites themselves, the tracking system, planning and development of facilities, and management of fiscal affairs. Each individual member nation, however, operates its own earth facilities, including transmitters, receivers, and antennas. While in principle all members have equal access to the system and are charged the same rate for the same service, INTELSAT does not control the rates a transmitting country may charge a receiving country for its signals. In addition to international satellite service, INTELSAT provides transponder space for domestic distribution on its satellites, a cost-saving measure that has led to leasing of space by wealthy countries such as the United States and Japan to poor countries such as Chad, Ethiopia, and Nigeria. As INTELSAT grew it added a number of special services. These include its Assistance and Development Program (IADP), which provides help mainly to developing countries in planning and establishing both international and domestic satellite services. INTELSAT also operates Project Access, which provides free access to UN-sponsored events aimed at improving health and education worldwide, such as the international AIDS concerts.

INMARSAT

Of an international nature, but specifically oriented to a particular class of users, is the International Maritime Satellite Organization (INMARSAT). Originally established to serve ships at sea, reminiscent of the early development of radio for the same purpose, INMARSAT expanded to serve the safety needs of travelers on land and in the air as well.

INTERSPUTNIK

Although operated for and by the eastern Europe and Asian nations that comprised the Soviet Union, INTERSPUTNIK membership is open to any country anywhere in the world. INTERSPUTNIK's several satellites cover almost the entire globe with video, audio and data.

ARABSAT

A number of regions operate satellite systems generally limited to the countries in their area. One such organization is the Arab Satellite Communications Organization (ARABSAT), which was founded in 1976. It has all twenty-two mideast Arab nations as members. However, relatively few of its member countries have ground stations with which to receive and distribute the signals, and in the early 1990s ARABSAT resources had not yet been fully used. More important factors, however, are the political differences among factions and individual Arab countries, which create a reluctance to admit visual or aural material from other countries. Ostensibly, the purposes of ARABSAT are of potential value to all Middle East countries: to provide area-wide educational programming; emergency communications in case of natural or artificial disasters; medical and health information and training; intergovernmental exchange of information and data; and on-line services such as linking libraries.

EUTELSAT

The European Telecommunications Satellite Organization (EUTELSAT) was established in 1977 to provide all European countries with their own cost-efficient means of distribution. It is the

**FIGURE 7-2
System map of
INTELSAT. Cour-
tesy INTELSAT.**

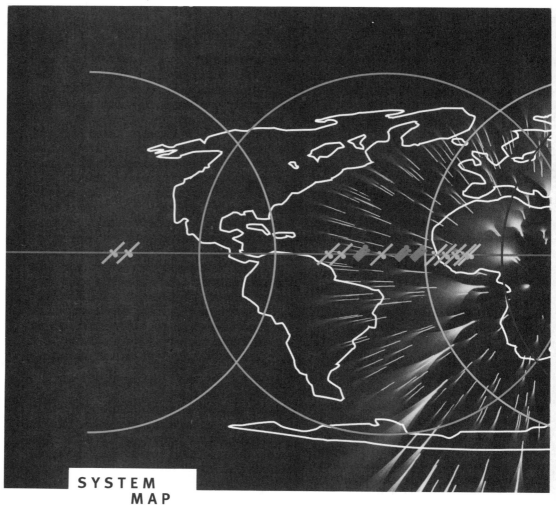

SYSTEM
MAP

System Highlights

- Nineteen spacecraft currently in service:
13 INTELSAT V/V-A satellites;
5 INTELSAT VI satellites;
1 INTELSAT K satellite.

- Eleven more spacecraft on order:
6 INTELSAT VII satellites;
3 INTELSAT VII-A satellites;
2 INTELSAT VIII satellites.

- Options for the use of three additional
spacecraft: Russia's INFORMKOSMOS
Express satellites.

- Four operating regions:
Atlantic
Indian
Pacific
Asia-Pacific

- 99.99 percent space segment reliability

- Maximum network flexibility and diversity
through a range of earth station choices.

- Six telemetry, tracking, control and monitoring
(TTC&M) stations to ensure the integrity of satellite operations and communications services.

FIGURE 7-2
Continued

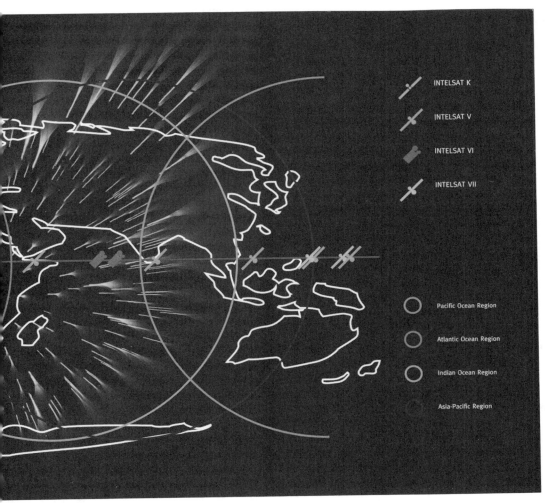

INTELSAT K

INTELSAT V

INTELSAT VI

INTELSAT VII

Pacific Ocean Region

Atlantic Ocean Region

Indian Ocean Region

Asia-Pacific Region

The INTELSAT System 1 June 1993

Atlantic Ocean Region

307°E	INTELSAT 513
310°E	INTELSAT 506
328.6°E	INTELSAT 504
325.5°E	INTELSAT 603
332.5°E	INTELSAT 601
335.5°E	INTELSAT 605
338.5°E	INTELSAT K
338.7°E	INTELSAT 502
342°E	INTELSAT 515
359°E	INTELSAT 512

Indian Ocean Region

57°E	INTELSAT 507
60°E	INTELSAT 604
63°E	INTELSAT 602
66°E	INTELSAT 505

Asia-Pacific Region

91.5°E	INTELSAT 501

Pacific Ocean Region

174°E	INTELSAT 510
177°E	INTELSAT 511
180°E	INTELSAT 508
183°E	INTELSAT 503

Future Satellite Deployments
(As of 1 June 1993)

INTELSAT 701	174°E
INTELSAT 702	359°E
INTELSAT 703	177°E
INTELSAT 704	66°E
INTELSAT 705	328.5°E
INTELSAT 706	307°E
INTELSAT 707	342°E
INTELSAT 708	57°E
INTELSAT 709	338.5°E
INTELSAT 801	174°E
INTELSAT 802	177°E

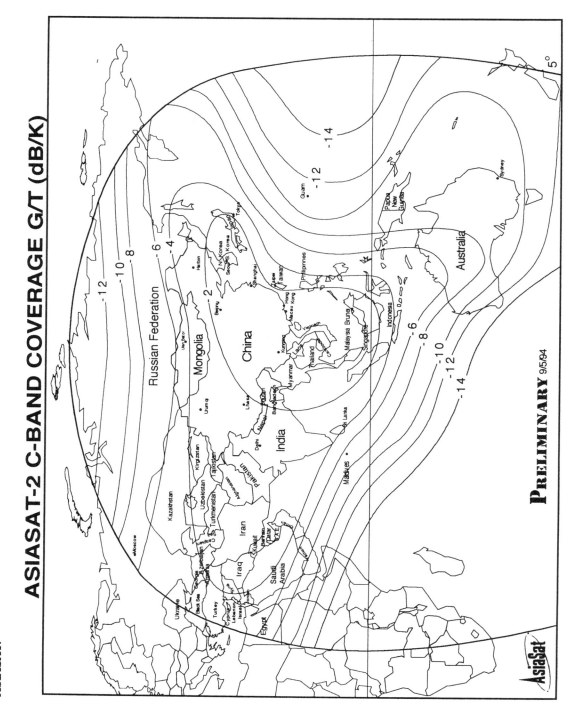

FIGURE 7-3
ASIASAT 2 cover-
age area. Courtesy
ASIASAT.

largest operator of satellites in Europe. EUTELSAT is similar to INTELSAT in organization and operation, except that its authority is limited to Europe. The European Space Agency (ESA) is a cooperative organization that coordinates satellite use, including the placement of EURTELSAT's satellites into orbit.

SES

The Societe Europeenne Des Satellites operates the ASTRA satellites. This pan-European communication service was launched in December 1988 and has become known as Europe's "Hot-Bird" for direct-to-home satellite signal reception.

Other Satellite Systems

In the planning stage in the early 1990s was the Andean Telecommunication Satellite System (SATS) to serve Latin America, the Regional African Satellite Communication System (RASCOM) to serve the entire African continent, and INSAT 1D and INSAT II to bring additional satellite service to India. Other regional satellite groups include ASIASAT for Asian countries (Figure 7-3),

CONDOR for South American countries, and CARIBSAT for Caribbean nations. Among domestic or national satellite services are DFS KOPERNICUS for Germany, TELECOM for France, HISPASAT for Spain, AUSSAT for Australia, BRAZILSAT for Brazil, GALAXY and SATCOM for the United States, ANIK E2 for Canada, and MORELOS and PANAMSAT for Mexico (Figures 7-4 and 7-5).

Ebert Foundation

While not an organization or association like those above, the Friedrich Ebert Foundation should be noted for its support to a number of cooperative groups for the principal purposes of news exchange and training workshops for broadcasting personnel. Ebert support goes to CARIBVISION, AFRIVISION, ASIAVISION, URTNA, and PACBROAD (Pacific Basin), among others.

Programming Groups

Europe

Although its name is the European Broadcasting Union (EBU), the EBU has a membership of and serves countries all

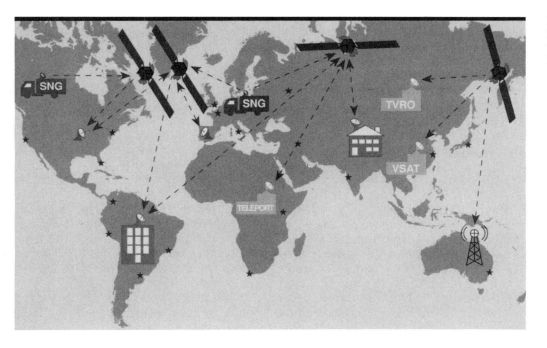

FIGURE 7-4
PanAmSat covers the globe. Courtesy PanAmSat.

FIGURE 7-5
Technical data from
PanAmSat media
kit. Courtesy
PanAmSat.

PanAmSat Satellites

Manufacturer: Hughes Aircraft Company		Satellite Bus: HS 601	
		Ku-band	**C-band**
Transponders:		16 x 54 MHz	16 x 54 MHz
Cross-Strapping Capability:		Up to 8 Transponders Ku to C	Up to 8 Transponders C to Ku
Transponder Output Power:		63 Watt	34 Watt
Redundancy:		HPAs 10:8 Receivers 4:2	HPSs 10:8 Receivers 4:2
Beams:		3 Uplink 5 Downlink	3 Uplink 5 Downlink
Satellite	**Ocean Region**	**Orbital Location**	
PAS -1*	Atlantic	45° WL / 315° EL (in operation)	
PAS-2	Pacific	192° WL / 168° EL	
PAS-3	Atlantic	43° WL / 317° EL	
PAS-4	Indian	292°/ 288° WL / 68°/ 72° EL	

*GE Astro Series 3000

over the world (Figures 7-6 and 7-7). The EBU developed out of political and military strife dating back to 1925, when the first of its predecessor organizations, the International Broadcasting Union (IBU) was formed. The beginning of World War II ended the IBU. In 1946 the Organization Internationale de Radiodiffusion, or the International Organization of Radio and Television (OIRT), was established, but the cold war resulted in the OIRT becoming principally an eastern European organization. Finally, in 1950 broadcasters in western Europe established the EBU. A nongovernmental organization, the EBU welcomes the membership of broadcast systems that serve an entire country. Its active members must be European or on the Mediterranean Sea. However, it offers associate membership to other countries from all over the globe. As a membership organization, it tries to avoid domination by any bloc and limits membership to two entities per country, but offers additional entities from any coun-

try limited privileges as "supplementary active" members. One of its principal purposes is to provide news programming to its members with greater objectivity than that provided by news services of countries with vested interests, such as BBC news and CNN news. In 1993 the EBU inaugurated a twenty-four-hour satellite news channel, Euronews. It also provides many other services, such as sports and entertainment programming. A Technical Division helps members in the development of ground and satellite networks. Its Legal Division represents the EBU membership in organizations and conferences on the international level, and also negotiates agreements for its members on such matters as copyright, acquisition of programs of international interest, coverage of live events such as the Olympics, and agreements with satellite and cable distributors.

In addition to providing news programs to its members, the EBU facilitates program exchange, offering special

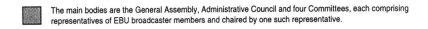

August 1993

STRUCTURE OF EBU

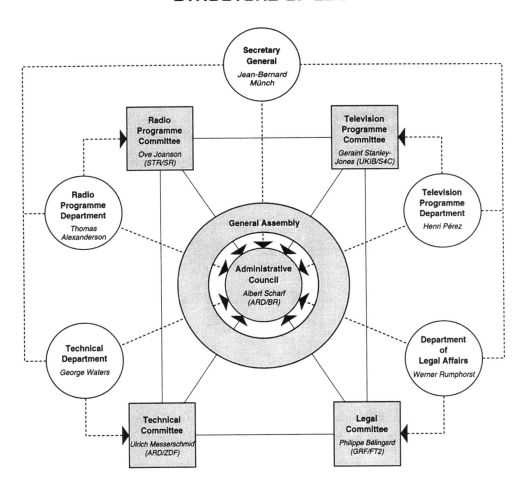

The main bodies are the General Assembly, Administrative Council and four Committees, each comprising
representatives of EBU broadcaster members and chaired by one such representative.

The Permanent Services, headquartered in Geneva, are headed by the Secretary General.
There are four departments corresponding to the four Committees.

FIGURE 7-7
EBU statistics.
Courtesy EBU.

- 118 members in 80 countries (64 active members in 48 countries and 54 associate members in 32 countries)

- 1993 budget: 360 million Swiss francs (including 170 million for rights acquisition and production costs, and 114 million for vision and sound circuit use)

- 230 members of staff employed at the Geneva headquarters, which welcomes 5,000 visitors a year

- in television, six satellite channels and 8.870 km of terrestrial circuits route 45,000 transmissions per annum (including 19,000 news items and 3,500 hours of sports and arts programmes)

- in radio, two satellite channels ensure relay of 1,500 concerts and operas, 400 sports events and 30 major news events

- in the technical and research spheres, 50 groups of experts produce one million pages of documents each year.

assistance as a clearinghouse in this regard to smaller countries that want quality programming but cannot afford to produce much of their own. Program exchange is done under the EBU's Eurovision. One of its projects is Eurosport, in which it participates jointly with SKY Television, headquartered in London, in a sports network. Its size and prestige enable it to obtain the rights to programs that its individual members might not be able to secure. For example, since 1960 it has obtained the broadcast rights for Europe for every Olympics.

It also serves as a clearinghouse for the exchange of information on matters relating to broadcasting. It works with new technologies and helps its members establish technical standards. Because of the different systems used by its members, it has been active in the development of converters and transcoders.

In the early 1990s, with the cold war ended and the velvet revolution in eastern Europe achieved, the EBU was negotiating with OIRT for a merger, with OIRT integrating into EBU to provide an even stronger organization. In 1991 Pierre Brunel-Lantenac, then Director of the EBU's News Development and Services section, told us of his increasing

efforts to provide news to and receive it from the former Soviet bloc countries in eastern Europe, with the ultimate aim of creating a free-flow program interchange system for all of Europe.

A number of small, private organizations have been established in Europe to serve specific needs. Some examples: MEDIA, which stands for Measures to Encourage the Development of an Audiovisual Industry, is directed primarily at smaller, less affluent countries. BABEL is the acronym for Broadcasters Across the Barriers of European Languages, which provides dubbing and subtitling services, with emphasis on the lesser-used languages. The European Film Distribution Office (EFDO) helps smaller countries in Europe with low-budget production. European Audio Visual Entrepreneurs (EAVE) offers training to professionals in the poorer countries.

Other organizations in Europe that generate impact on telecommunications include the Conference of European Postal and Telecommunications Administrations (CEPT), which has periodic conferences to deal with satellites, transatlantic telecommunications, data, and other areas relating to their responsibilities as government agencies; the

European Telecommunications Network Operators (ETNO), an offshoot of CEPT, representing the interests of the telecommunications carriers; and the European Institute for Research and Strategy Studies in Communication (EURESCON).

Finally, it is important to mention again the European Union and its impact on telecommunications, specifically through its open border policies that promote and facilitate exchange of programs and co-production among its members, and through its broadcasting directive that calls for a majority of programming time in any and every member country to be reserved for European works. The Directive also affects content, as discussed earlier, especially in relation to limits on advertising time and placement, and certain products, such as drugs and tobacco, and in relation to material that "might seriously impair the physical, mental or moral development of minors, in particular those that involve pornography or gratuitous violence," and to "ensure that broadcasts do not contain any incitement to hatred on grounds of race, sex, religion or nationality."

Africa

Africa's principal telecommunications coordinating group is the Union of National Radio and Television Organizations of Africa (URTNA). While its stated purposes target the principal needs of the region, it has been faulted for not following through on its potentials and commitments. Among its aims in promoting, developing, and coordinating both radio and television throughout Africa are the exchange of programming, the establishment of co-production and other cooperative projects, and the strengthening of African culture through an emphasis on local production as opposed to foreign imports. URTNA has from time to time carried out studies of the video production industry in Africa, held seminars on the place of broadcasting within national priorities, and offered training programs for the operation and management of broadcast stations. In the early 1990s, as television broadcasting and satellite use grew among its forty-eight member countries, URTNA increased its activity in assisting the development of radio and TV. However, it is expected that the satellite coordinating group, RASCOM, will in the future play the key role in coordinating African telecommunications operations, with a link to INTELSAT. Subregional groups, as mentioned earlier, began to develop in the mid-1990s for program and technical cooperation among adjacent countries.

In 1994 AT&T announced a highly imaginative plan for communications distribution throughout Africa. It proposed to put a fiber-optic ring around the entire coast of Africa, with headends fed by satellite, traffic to all of Africa through the communications ring, and internal distribution via microwave.

Arabia

The Arab States Broadcasting Union (ASBU) objectives are to coordinate program exchange, including news; arrange for area-wide distribution of international events, such as the Olympics; distribute programming, especially news, from its member states to other regions and countries globally; provide training seminars and workshops; and represent its members in relation to international and other regional telecommunications organizations. Another group, discussed under satellite organizations, is ARABSAT, which by the nature of its operations serves as a program exchange and distribution center, as well as a technical coordinator. ARABSAT has already provided, additionally, the base for area-wide cooperation, training for broadcast professionals, and region-wide news gathering and dissemination. ARABSAT and ASBU serve countries in both Africa and Asia.

Asia

The most important telecommunications organization in Asia is the Asia-

ASIA-PACIFIC BROADCASTING UNION

LIST OF MEMBERS

FULL MEMBERS
(38 members in 30 countries)

Radio-Television of Afghanistan (RTA)
Australian Broadcasting Corporation (ABC)
Special Broadcasting Service Corporation, Australia (SBSA)
National Broadcasting Authority of Bangladesh (NBAB)
Radio Television Brunei (RTB)
Radio and Television of the People's Republic of China (RTPRC)
Egyptian Radio and Television Union (ERTU)
Fiji Broadcasting Commission (FBC)
All India Radio (AIR)
Doordarshan, India (DDI)
Radio Republik Indonesia/Televisi Republik Indonesia (RRI/TVRI)
Islamic Republic of Iran Broadcasting (IRIB)
Iraqi Broadcasting and Television Establishment (IBTE)
Nippon Hoso Kyokai, Japan (NHK)
Tokyo Broadcasting System, Inc. (TBSJ)
Jordan Radio and Television (JRTV)
Korean Broadcasting System, Republic of Korea (KBS)
Lao National Radio and Television (RTNL)
Radio Television Malaysia (RTM)
Department of Information and Broadcasting, Maldives (DIB)
Nepal Television Corporation (NTV)
Radio Nepal (RNE)
Television New Zealand Limited (TVNZ)
Pakistan Broadcasting Corporation (PBC)
Pakistan Television Corporation Ltd. (PTV)
National Broadcasting Commission of Papua New Guinea (NBC/PNG)
Philippine Broadcasting Service/People's Television 4 (PBS/PTV4)
Republic Broadcasting System, Philippines (GMA)
Saudi Arabian Radio and Television (SAB/SAR)
Singapore Broadcasting Corporation (SBC)
Solomon Islands Broadcasting Corporation (SIBC)
Sri Lanka Broadcasting Corporation (SLBC)
Sri Lanka Rupavahini Corporation (SLRC)
National Broadcasting Services of Thailand (NBT)
TV Pool of Thailand (TPT)
Turkish Radio-Television Corporation (TRT)
Ministry of Culture and Information, Socialist Republic of Vietnam (VNRT)
Samoa Broadcasting Service, Western Samoa (SBS)

ADDITIONAL FULL MEMBERS
(22 members in 12 countries and areas)

Asia Television Ltd., Hong Kong (ATV)
Hong Kong Commercial Broadcasting Co., Ltd. (HKCBC)
Metro Broadcast Corporation Ltd., Hong Kong (METRO)
Radio Television Hong Kong (RTHK)
Television Broadcasts Ltd., Hong Kong (TVB)
Pt. Rajawali Citra Televisi Indonesia/Pt. Surya Citra Televisi (RCTI/SCTV)
Pt. Cipta Televisi Pendidikan Indonesia (TPI)

Asahi National Broadcasting Co., Ltd., Japan (ANB)
Christian Broadcasting System, Republic of Korea (CBSK)
The Radio and Television Broadcasting Committee of the Democratic People's Republic of Korea (KRT)
Munhwa Broadcasting Corporation, Republic of Korea (MBC)
Seoul Broadcasting System, Republic of Korea (SBSK)
Teledifusao de Macau, S.A.R.L. (TDM)
Sistem Televisyen Malaysia Berhad (TV3)
Mongol Radio and Television (MRTV)
Norfolk Island Broadcasting Service (VL2NI)
Shalimar Television Network Ltd., Pakistan (STN)
ABS-CBN Broadcasting Corporation, Philippines (ABS-CBN)
Radio Philippines Network, Inc. (RPN)
Independent Television Network Limited, Sri Lanka (ITN)
MTV Channel (Pte.) Ltd., Sri Lanka (MTV)
Telshan Network (Pte.) Ltd., Sri Lanka (TNL)

ASSOCIATE MEMBERS
(28 members in 20 countries and areas)

Federation of Australian Commercial Television Stations (FACTS)
Nine Network Australia Limited (NNA)
Canadian Broadcasting Corporation (CBC)
Finnish Broadcasting Company Ltd. (YLE)
Organismes Francais de Radio et de Television, France (OFRT)
A.R.D., Republic of Germany (ARD)
Zweites Deutsches Fernsehen, Republic of Germany (ZDF)
Radio Telefis Eireann, Ireland (RTE)
The National Association of Commercial Broadcasters in Japan (NAB)
Radio Kiribati (BPA)
Mauritius Broadcasting Company (M$_A$BC)
Radio New Zealand Limited (RNZ)
Federated States of Micronesia Broadcasting Service (FSMBS)
Radio Netherlands (RN)
Federal Radio Corporation of Nigeria (FRCN)
Norwegian Broadcasting Corporation (NRK)
Islands Broadcast Corporation, Philippines (IBC)
Radiotelevisao Portuguesa, E.P. (RTP)
Sveriges Radio AB, Sweden (SR)
Swiss Broadcasting Corporation (SwBC)
British Broadcasting Corporation (BBC)
American Broadcasting Companies, Inc., USA (ABC-USA)
Columbia Broadcasting System, Inc., USA (CBS-USA)
Corporation for Public Broadcasting/Public Broadcasting Service/National Public Radio/American Public Radio Network (CPB/PBS/NPR/APRN-USA)
National Association of Broadcasters, USA (NAB-USA)
National Broadcasting Company, Inc., USA (NBC-USA)
Turner Broadcasting System, Inc., USA (TBS-USA)
Voice of America (VOA-USA)

TOTAL: 88 members in 51 countries and areas.

FIGURE 7-8
APBU's membership list.
Courtesy APBU.

Pacific Broadcasting Union (ABU) (Figure 7-8). It reaches beyond Asia with members from Oceania, principally Australia, and with associate memberships from some other countries throughout the world, including the United States and the United Kingdom. The ABU's functions are similar to those of other regional groups: to promote and facilitate program exchange, to coordinate exportation and importation of programs, to encourage co-production and other cooperative projects among member nations, and to provide training and technical assistance, especially for developing countries. ABU's functions also includes, importantly, the promotion and distribution of educational programs, including formal instruction, especially to remote areas where educational opportunities are not otherwise available.

ABU pays special attention to news, promoting and facilitating interregional news exchange, and holding workshops for journalists and program directors. One of its aims is to reduce the political, legal, and technical barriers among countries that frequently tend to reduce the flow of news. It cooperates with two other organizations in Asia, one concentrating on the news, the Asian Pacific News Network, and the other oriented to the growth of telecommunications, including programming, the Asia-Pacific Institute for Broadcast Development.

With so many of its countries in a "developing-nation" status, various efforts have been made in Asia to provide not only technical and monetary assistance in the development of telecommunications systems, but also education and training so the systems can be effectively and efficiently operated. The Asian Institute for Broadcasting Development (AIBD) supports regional centers offering technical education. The Asian Center for Educational Innovation and Development (ACEID) promotes the inclusion of broadcasting studies in college and university curricula. The Asia Mass Communication Information and Research Center (AMIC), in Singapore, is a clearinghouse for materials relating to the study of mass media in Asia, and promotes the development of telecommunications throughout the region. It has a counterpart in South Korea, the Asian Mass Communication and Research Center, at which one of this book's authors lectured in 1992 and was impressed by the strong support by noncommunications political and professional leaders for the study and development of broadcasting in the area. Additionally, the Chinese Radio and Television Society, located in Beijing, promotes the development of the broadcasting sciences and seeks to maintain the quality of electronic media in the People's Republic of China.

Another important organization is the Southeast Asian Ministers of Educational Organizations (SEAMCO). Recognizing the significance and, in many areas, the crucial importance of radio and television in providing equitable and quality education, SEAMCO educates personnel from its southeast Asia member states in the development and use of instructional materials on radio and TV.

Oceania

The Pacific Basin countries, also frequently referred to as Australasia, include Australia, New Zealand, Fiji, Papua New Guinea, and other island nations in the region. Most of these nations belong to the Asia Pacific Broadcasting Union, which provides essentially the same type of coordinating and facilitating services as do similar unions in other regions of the world. However, with Australia the dominant telecommunications nations in the area by far, and with all the other nations, with the major exceptions of New Zealand and the Philippines, dependent on Australian communications, all regional organizations are not only oriented to but dominated by what Australia can and wishes to offer. Even the regional satellite system is called AUSSAT. Most of the countries in the area participate in the ABU. The South Pacific Telecommunications Programme was established by the South Pacific Economic Commission to assist in the development and coordination of telecommunications facilities and programming in the region.

INTERNATIONAL ASSOCIATIONS

A number of professional and trade associations attract international membership. These members are usually individuals, institutions, and businesses that seek contacts, information, sales, and means for joint research, projects, and ventures. While these are not official bodies consisting of government members, they do represent grass-roots international cooperation by individuals from various countries who meet, exchange, and collaborate. One of the most important of these groups, of special significance to educators and students reading this book, is the International Association for Mass Communication Research (IAMCR), whose membership is composed primarily of scholars and professors from corporations, educational institutions, foundations, think tanks, and government offices from countries all over the world.

The IAMCR divisions, representing the kinds of topics under study—and in some cases projects and solution implementation—in many parts of the globe, are: communication technology policy; documentation and information systems; gender and communication; communication history; international communication issues; international communication law; local radio and television; media education research; political communication research; political economy; professional education; and sociology and social psychology aspects of communication. The IAMCR, like some other international communication associations, dedicates considerable effort to the role of communication in achieving peace and human rights.

The International Communication Association (ICA) is similar to the IAMCR in that it consists of individual and group members from all parts of the globe, holds conferences, and conducts projects and studies relating to various aspects of world communications.

Other international organizations worth noting include the Commonwealth Broadcasting Association (CBA), the Committee on the Peaceful Uses of Outer Space (COPUOS), the International Association of Broadcasting

FIGURE 7-9
Senegal was the site
of AMARC's sixth
world gathering.
Courtesy AMARC.

(IAR-AIR), and the World Association of Community Broadcasters (AMARC) (Figure 7-9).

Global Equality

Though the increasing emphasis in recent years on human rights and equal opportunity in many countries has permitted the entry of increasing numbers of women and members of previously barred racial and ethnic groups into telecommunications fields, the leaders and managers in the telecommunications industries have continued to discriminate strongly against women and these other groups. As a consequence of both the presence of women and others in the field, and the continuing biases against them, a number of national and some regional and international associations of representatives of these groups have been formed. Perhaps the most extensive affiliations to push for greater equality of opportunity in worldwide telecommunications are those of women.

The oldest and largest such organization is the International Association of Women in Radio and Television (IAWRT), which was established in 1950. Twenty-seven countries are represented in the IAWRT—a large enough number considering that many countries of the world, especially in Africa and Asia, consider and treat women as chattel rather than as human beings, and under pain of punishment and even death women in some of these countries are forbidden to become involved in telecommunications. Even in many countries where they are allowed some access, they are forbidden to participate in organizations that might be considered of a politically active nature. The IAWRT receives support from the United Nations Development Fund for Women, and has as a principal aim assisting women to enter or advance in telecommunications in countries everywhere. It puts great emphasis on third-world countries, in most of which women face especially difficult, if not insurmountable, barriers. One of the IAWRT's publications is a listing of women broadcasters who offer their personal assistance to women in developing nations. The IAWRT also attempts to make known women's points of view, which are largely ignored by most media throughout the world, in critical areas such as peace, human rights, and individual and national growth.

In Africa the Federation of African Media Women (FAMW) is a resource center for information and assistance for women in the media on that continent. One of its purposes is to create more objective and less prejudicial portrayals of women in the media. It also attempts to initiate media programming aimed at improving the conditions of women and others in the region. It was founded in 1981.

The Women and Media Network for Asia and the Pacific (WMNAC) was established in 1988 with the help of UNESCO. Because of the geography of the region, its work is decentralized. It distributes information for and about women in the media and provides training programs for women in the field. WMNAC works closely with the Asian-Pacific Institute for Broadcasting Development (AIBD), mentioned earlier.

On the heels of the dissolution of the Soviet Union arose Independent Broadcasting for Women, headquartered in Moscow. The IBW's mission is the promotion of women in the broadcast arts in the new Russia.

Appendix to Chapter 7

Communication scholars Nancy Lynch Street and Marilyn J. Matelski provide an account of their experiences while researching the United States' external services for their planned book on the fate of the "Radios" in the post-Soviet bloc era. This essay was presented at the Eastern Communication Association's April 1994 conference.

"EUROPE IS FREE— WHO NEEDS FREE EUROPE?"[1] FEDERALIZING SURROGATE RADIO

Nancy Lynch Street

Bridgewater State College

Marilyn J. Matelski

Boston College

Perhaps one of the most important aspects of the communication system of the Western democracies (however different the individual approaches) is the media system, including all forms of media and its practitioners—the working journalists, some of whom, perhaps many, believe that "the truth shall make ye free." This then can be viewed as a segment of the ideological infrastructure necessary to the working Western journalist. The question then becomes: if the equation is truth plus action equals freedom—what then? Does democracy necessarily follow? In the past year, this has become a central question in the battle to phase out Radio Free Europe/Radio Liberty (RFE/RL) as an independent international news organization, and to merge it with other federal information agencies, which include the Voice of America (VOA), Radio Marti, Worldnet and Radio Free Asia.

Our first direct encounter with Radio Free Europe/Radio Liberty as a serious influence upon the democratization of the Soviet Bloc occurred in August 1990, while researching an intercultural project in Eastern Europe. In Poland, we visited the Shrine of the Black Madonna at the Jasna Gora Monastery in Czestochowa, arguably the most sacred site in Poland. Following the viewing of the icon of the Black Madonna, we were surprised to find ourselves in a room filled with Solidarity memorabilia (missing during a visit in 1993), heightening our awareness of the partnership between the Church and the media in the triumph of the Solidarity Movement in Poland. Our monk-guide quickly confirmed the crucial partnership between the Church and the (underground) media in the ferment boiling in Hungary and Poland in the late eighties. We left the Shrine, convinced that the next stage in our study of the dissolution of the Soviet Union should be focused on discovering the role of transnational radio (Vatican Radio, Radio Free Europe/ Radio Liberty, the Voice of America and the British Broadcasting Corporation) in the "end of the Cold War," symbolized by the dissolution of the Iron Curtain and the fall of the Berlin Wall.[2]

Intrigued by the possibilities raised by our revelation at Jasna Gora, we decided in the Fall of 1992 to research the roles of the American Radios of both VOA and RFE/RL at their broadcast sites in Munich, Germany,[3] not realizing that by spring 1993, all three Radios would be under intense scrutiny as the budget and deficit issues loomed large on the American national agenda. Briefly synthesized, VOA and RFE/RL were fighting for life, dollars—and who should have power.

In Fall 1992,[4] as we were beginning our exploration of transnational radio and the demise of the Soviet Bloc, George Bush was living out (what turned out to be) his last days as America's commander-in-chief. Management and personnel at Radio Free Europe and Radio Liberty, however, may have had a more vested interest in the 1992 presidential race than most; for the fate of the world's largest transnational surrogate radio service seemed to depend heavily upon the man chosen to be the next president.

Despite having felt some pressure in 1990 during the Bush administration, the Munich-based, privately funded Radio Free Europe/Radio Liberty (also known as RFE/RL, or "the Radios") had survived intense scrutiny by a special commission for international broadcasting, the Hughes Commission, deployed by President Bush in 1991. The Commission, specifically charged with devising a plan to consolidate all international broadcasting by the United States, concluded (after six to seven months of research-gathering and public hearings) that the various international broadcasting institutions, i. e., RFE/RL, VOA (Voice of America), the Office of Cuba Broadcasting (including Radio Marti) and the proposed Radio Free Asia, should not be consolidated.[5] Further, the Commission's findings indicated that Radio Free Europe would be needed for the next few years and that Radio Liberty should continue well into the next century. Unfortunately, according to Gene Pell (former President of the Radios),[6] this report went largely ignored "because it didn't fit the conventional wisdom of some who thought the outcome should be different."

In fact, the goals and objectives of each of the transnational broadcast systems identified in the Hughes Commission Reports differ markedly. Radio Marti and the proposed Radio Free Asia, for example, have very specific geographic audiences, issues and political concerns. Radio Free Europe/Radio Liberty, on the other hand, recognize regional differences, but are most well-known for their unique approach to journalistic reporting. As stated in their original mission statements, the Radios strive to be "surrogate broadcasters"—providing unfettered news to nations who have little or no access to it. This programming philosophy contrasts sharply from the Voice of America, which exists to project the United States government's position, policy and image throughout the world in English and other languages.

With this in mind, we set about to travel to Munich—the center of growing transnational broadcast controversy—stopping along the way to interview people in Eastern Europe, and most specifically in Poland,[7] about their beliefs and attitudes toward VOA and the Radios.

By the time we arrived in Munich, having first travelled by German and Polish railway from Berlin via Warsaw to Gdansk and talked with leaders of Solidarity—regarding their impact on Poland from 1980 to 1989, as well as the current status of Solidarity in Poland under the new government[8]— the Munich situation had undergone rather dramatic revision. In the few days we spent in Krakow, we began to understand not only some of the primary issues associated with the political and economic evolution of the Polish culture and economy ($100 American is worth approximately 1,800,000 zlotys)—but also some of the problems associated with the continuing independent operation of RFE/RL in especially Poland, the newly-founded Czech Republic (the result of the division of Czechoslovakia into the Czech Republic and Slovakia) and Hungary.

We arrived in Munich prepared to spend a week at RFE/RL and VOA. At the beginning of the week the staff in Public Affairs thought that the die would be cast and that by the end of the week the Clinton proposal to merge RFE/RL and the other international services under the USIA Director and a newly-created Board of Governors would be finished. This was the week of August 2; we left Munich for Berlin on August 7. The final decision was, in fact, made in October.

On October 18, 1993, Eugene Pell finally voiced the words his staff had been dreading for over two years: By 1996, more than 800 positions would be eliminated from the Radio Free Europe (RFE) service. As this announcement reverberated throughout the long corridors of its Munich-based headquarters, Pell's earlier warnings about RFE's ultimate demise became real—since all of Europe was now "free," there apparently would be little need for "Free Europe." For those who support the principles of surrogate broadcasting, i.e., speaking for those who have no voice, the proposed elimination of this respected, transnational radio service was premature and ill-advised.

In the meantime, the staff at RFE/RL continued to transmit (from within each country, as opposed to relying on stringers living abroad)[9] news and features such as RFE/RL's "The Democratic Experience" and RL's Moscow-based "Face to Face," as well as an ambitious live, two-way broadcast between Munich and Moscow entitled "Liberty Live." During the time we spent at RFE/RL headquarters (located near the English Gardens, about two blocks from the Munich Hilton) the Washington decision regarding federalization of the service was of primary concern both at RFE/RL and at the VOA, which has its European offices in Munich at the offices of the United States Information Agency, to which it reports.

From our interviews with the management staff at the Radios, including Gene Pell, William Marsh (RFE/RL Executive Vice President, Programs and Policy), Kevin Klose (Director, Radio Liberty) and Robert Gillette (Director, Radio Free Europe), a clear distinction was drawn between the Radios and the Voice of America. To compare the Radios and their missions, VOA, founded in 1941, is a world-wide service, broadcasting partly in English and partly in other languages. Like CNN, VOA-Europe broadcasts in English only and its "programming is based upon scripts centrally written by Americans." VOA's mission, according to an editorial in *The Wall Street Journal*,[10] is to "describe America to the outside world." Or, as Walter Laqueur writes in *The Wall Street Journal*,

There is no overlap [between RFE/RL and VOA]. The task of VOA is, to put it inelegantly, to "sell America." . . . VOA has many merits, but its direct political impact in Russia and Eastern Europe is almost nil, whereas that of the Munich radios is immense.[11]

VOA is not only generally perceived to be a voice of the U.S. government, it is *in fact* a voice of the U.S. government, headquartered in Washington, D.C. No one disputes that it does its job well. But there are those who, particularly in Eastern Europe and the former Soviet Union, fear and distrust the VOA as a government mouthpiece. Given the 40–70 years history of government controlled media, small wonder that there is a distrust of any media speaking for any government. As American citizens have reason to know, in our own history, we are, from time to time, wary of the media. For the former Soviet Union and Eastern European countries, the experience is still vivid. Lech Walesa, former Solidarity leader and now president of Poland, wrote in 1993:

There are those institutions which have changed our world. Without them, the world would have been completely different. The Polish Section of Radio Free Europe is one such institution. . . . How fortunate that "the iron curtain" could not be raised so high as to block radio transmissions. Although the boundaries were impenetrable for people, they were not so for shortwaves. The truth seeped in, unseen by the border guards and their dogs, above the mine fields, between the barbed wire, alongside reconnaissance planes and patrol boats. It proved impossible to stop; impossible to silence it."[12]

From broadcast content to home base and lifestyle VOA is American and, unlike RFE/RL, it was not meant to be a surrogate facility, nor has it evolved over time and under changing conditions, into a surrogate news facility. Despite the rationale developed for implementation of the proposed plan, VOA and RFE/RL do not overlap in programming. The issue is two-pronged—money and the "vision thing"—and the United States, with its traditional respect for freedom of speech and the role of the media in maintaining that tradition within democracies, seems to have lost sight of the difference between propaganda and news. As Kevin Klose, former Moscow correspondent for the *Washington Post* and present Radio Liberty Director, puts it,

Journalism, well done, is very powerful and very fragile, you start messing with it, its sense of self, its independence from the people who pay its way and things start changing in the cadres' head and pretty soon, very quickly, it's up the antenna and out. These are people {in the recently released countries of the former Soviet Union} who have spent the last 75 years struggling with the poison of government run media, they are very, very, sensitive to the issue of how it sounds, is it credible enough.[13]

By Wednesday, after two days of absorbing RFE/RL information, we set off for the offices of the VOA to catch their reaction to the proposed structure of international broadcasting. The casual visitor would have great trouble finding the VOA in Munich as no signs illuminate one's way through the maze of offices and no flags fly to indicate the American presence. Located on Ludwigstrasse, near one of the wealthiest shopping districts in Europe, the edifice looks more ambassadorial than journalistic. Bewildered about where to go next, we encountered a kindly cleaning woman who finally showed us the way.

The newly-appointed manager for affiliate relations, Ted Lipien (originally from Poland), answered our questions, offering no dispute when we shared RFE/RL perceptions of the proposed reorganization with him. In short, Lipien tended to confirm the accuracy of RFE/RL's perceptions regarding the federalization of the Radios.

Whatever its undisputed achievements, the Voice of America is still perceived as an official mouthpiece, transmitting official material and propaganda. In an editorial, Kazimierz Woycicki (Editor-in-Chief, *Zycie Warsawy*) describes his view of the role of RFE/RL:

Modern democracy is largely predicated on the existence of a proper order and civilized relations within the media as an element of the democratic system. . . . the Munich based radio station owes its enormous influence to the fact that it has created its unique style of journalism. RTE transmits, to a far greater degree, the democratic way of thinking and attitude towards the world.[14]

How and why are the Radios perceived differently? In contrast to the VOA, RFE/RL is currently, at least until 1995, responsible to the BIB or Board of International Broadcasting rather than to the USIA. When the Radios were founded (RFE in 1949; RL in 1951), they were not perceived favorably as they were funded by the US government through the CIA (there are still today at the Munich headquarters a few former CIA employees). Further, their policies and promises were not always in sync with the possible as during the Hungarian uprising, during which RFE/RL reportedly encouraged the Hungarian people by saying that help would come—and it did not. Chastened by this experience, the radios began to streamline their system, though there is no question that one major goal of their program was to encourage dissent, to bring about, if possible, the dissolution of the Soviet Union and Eastern Europe.

In the seventies, the Radios again came under fire during the Nixon administration and in 1971 "all connections with the CIA were severed"[15] and, in 1973, the BIB was formed. The two Radios merged as RFE/RL, Inc., on October 1, 1976. The budget for RFE/RL is allocated by Congress and is approximately 220 million dollars a year. Thus, since 1976 RFE/RL, unlike VOA, is a private entity which is not responsible to the United States government. The Radios "corporate existence" is therefore quite different. As mentioned previously, VOA is headquartered in Washington, while RFE/RL is presently located in Munich (soon to be re-established in Prague). This situation has considerable impact upon the hierarchical structure and the tax situation of each entity and of its individual employees as well as negotiation for AM/FM space with governments of nations receiving both VOA and RFE/RL signals.

In the furor created by the Clinton administration's proposal to dissolve RFE/RL, integrating or federalizing them, *Pravda* has commended the proposal, commenting bitterly on the role of RL in the dissolution of the Soviet Union. In Romania, Mircea M. Dabija, in *Europa*, writes:

Radio Liberty had and still has as its target the current area of the former Soviet Union, and its sister, Radio Free Europe—the area of the other former socialist countries from the center and east of our continent. These radios inspired, established and actually financed by the main intelligence agency of the USA, the CIA, in close connection with the USIA, came into existence at the beginning of the hot cold war, which they ignited and kept up industriously and fanatically, being some of the most important and effective propaganda means of the capitalist West against the socialist East.[16]

Meant as a stinging indictment of the radios, the Munich-based RFE/RL uses the article as evidence of its continuing success and effectiveness—the material can be found in the packet prepared for the Congressional delegation to Munich in the spring of 1993, during the budget hearings. The article does serve, however, to illuminate the complexity and ambiguity of RFE/RL's past and future mission, raising the question of the integrity of the existing Radios (and the proposed Radio Free Asia Service) and its ability to foster the ideal or, Western objective journalistic practices and ethics.

Relatedly, journalist Adam Michnik, an opponent of the Clinton plan summarizes the view of many:

In all these countries dictatorship has lost and freedom has won. But that does not mean that democracy has won. Democracy means the institutionalization of freedom. We don't have a democratic order, and that is why our freedom is so fragile and shaky.[17]

Michnik is making a clear distinction between freedom and democracy, illuminating the task ahead which is more difficult in the short term than the peoples of these newly-freed countries were prepared for. As Robert Pirsig says in *Zen and the Art of Motorcycle Maintenance*, you can change the system but if you don't change the system of thinking underlying it, nothing changes. However much one may chafe under an authoritarian regime, the reality is that one gets used to and absorbs the prevailing way of thinking, unknowingly. Further, as in the communist hierarchy, certain people manage, others obey. The only persons who know how to manage in formerly communist countries are—the old managers! Thus, there is a practical dearth of information/expertise on democracy and management in the former Soviet Union and in Eastern Europe. This situation, argue the journalists at the Munich-based RFE/RL, necessitates that RFE/RL not be absorbed into the Voice of America (VOA) as the new plan issued by the Board for International Broadcasting indicates will occur by the end of 1995. Simply put, RFE/RL claims that merging with VOA invalidates the political, social and linguistic credibility of RFE/RL which is based upon news gathered and delivered by nationals in the language of the country. Secondly, RFE/RL provides the training ground for Eastern European journalists, necessary to developing Western media integrity, perceived to be essential to the development and maintenance of democratic institutions. Third, RFE/RL argues that there is still a strong need for surrogate international broadcasting, providing the peoples of Eastern Europe and the Soviet Union with an objective, alternative view of their own societies, their neighbors and the world at large. . . . encourage the understanding and spread of democratic ideas, values and practices. . . . and provide a model of journalistic practices and ethics.[18]

There are two issues here. The first issue concerns the status of each Radio;

the second issue concerns the incredible technical and technological changes occurring within transnational media worldwide. The status and technological issues are connected. That is, VOA negotiates with foreign governments on a different footing than does RFE/RL for transmitter sites and access to AM and FM channels. As part of the United States government, VOA negotiations with other governments often are more costly and restrictive than those of RFE/RL, since the latter is a private corporation. Thus, if the status of RFE/RL should be altered, i.e., federalized, all contracts currently in place will be renegotiated in view of its new status. Therefore, one of the major reasons for federalization, to save money, will vanish as new costs accrue under the newly created design for American international broadcasting.

Yet another, more serious consideration is that of perception and credibility. Should RFE/RL be federalized, the perception of the Radios by other persons and governments will alter. Senator Biden states the dilemma well:

The issue is one of organization for this agreed purpose. In essence, can surrogate radios function effectively, with journalistic integrity and credibility, if their analysts and journalists are employees of the U. S. government?[19]

Here Senator Biden strikes a nerve center of journalistic principles. For years, Americans have railed against government controlled presses, advocating the free press approach followed in the United States. The media is conceived to be the "watchdog of Democracy." Why then are we proposing to impose yet more governmental control over other nations? If we know, as we do, that most of nations of the former Soviet Union and the Eastern European countries still have much the same archaic hierarchy-often run by former communists—and media infrastructure in place, now manned by poorly trained (by Western journalistic standards) journal-

ists, why not continue to offer them (as RFE/RL now does) an alternative voice and rigorous training in journalistic principles, practices and ethics? Why propose to introduce governmental control of the media—this time from the West, specifically from America? Is this akin to our policy of "dumping" America's cigarettes on second and third world countries without thought of the consequences? What of the consequences of losing a needed voice, particularly when virtually everyone but the U.S. Congress recognizes that while the Cold War may be over, democracy is not in place, and the newly freed countries are easy prey in perilous times for authoritarianism (or communism in new dress)? The view seems short-sighted—but when have we ever done the "vision thing"?

In Senator Biden's words,

It was a staple of the Cold War that Americans mocked countries that deployed "Journalists" in the employee [sic] of governments. It would be a nice but unpleasant irony were we to mark the end of the Cold War by adopting this practice ourselves. Can anyone actually argue that journalism and government employment are compatible?[20]

ENDNOTES

1. Radio Free Europe/Radio Liberty (RFE/RL) President Gene Pell characterizing the "soundbite" approach to the proposed dissolution of RFE/RL during an interview with him in his office in Munich, Germany, Tuesday 3 August 1993.
2. This "study" ultimately became more complex than we had originally planned. We now are writing a book on our research, tentatively entitled *Messages from the Underground: Transnational Radio in Resistance and in Solidarity.*
3. Today, Radio Free Europe/Radio Liberty's major facilities are still located in Munich (although plans are underway to move to Prague). The Voice of America's programming center is in Washington, D.C. However, VOA also has strong

historical ties to Munich, having used the city as its primary transmission site for much of its fifty-year history.

4. The development of this project, including rail travel in Europe and a research assistant, was aided by a CART Grant from Bridgewater State College in 1992–1993.

5. From the above interview with Gene Pell, President of RFE/RL. For testimony and findings, see the Hughes Commission Reports.

6. Mr. Pell announced his resignation from RFE/RL shortly after the announcement of the federalization plan. He left the Radios in October 1993.

7. We conducted several interviews in Krakow, one of which was with Bronislaw Wildstein who prior to the end of the Cold War was an RFE/RL stringer living in exile in Paris. Following the fall of the Berlin Wall, Wildstein was invited to return to Poland and became the Director of Radio Krakow. We met with him in July of 1993.

8. We met with Andrzej Matla, Head, International Department of Solidarity and others. No longer a social movement, Solidarity is once again a trade union and a force in the political dimension of Polish life. Further, it is not tied to President Lech Walesa. Any discussion in Poland or elsewhere of President Lech Walesa was at best ambiguous, at worst, hostile. One Polish respondent says, "as an electrician, he's a good president."

9. Prior to 1989, RFE/RL's broadcasts did not originate within the countries which they served. News and features were suggested/provided by the stringers (often dissidents, once in a while spies) who lived abroad in such major cities as London, Vienna and Paris. Now RFE/RL has bureaus in the countries which it serves.

10. Editorial, "Yeltsin Should Ask," *The Wall Street Journal* (April 2, 1993). It should be noted that this editorial (as well as all other incomplete cites) were taken from a Radio Free Europe/Radio Liberty Press packet given to us at the outset of our visit.

11. Walter Laqueur, "The Dangers of Radio Silence," *The Wall Street Journal* (March 4, 1993).

12. Lech Walesa, "From the President of the Republic of Poland, Warsaw, May 1992, to the Polish Section, Radio Free Europe," *Board for International Broadcasting 1993 Annual Report on Radio Free Europe/Radio Liberty*, Malcolm S. Forbes, Chairman, Washington, D.C.: Board for International Broadcasting, 1993: 5.

13. Kevin Klose, Director of Radio Liberty, in an interview at RFE/RL in Munich on 2 August 1993.

14. Kazimierz Woycicki, "Eastern Europe Still Needs this Message," *Zycie Warsawy*, n.d.

15. Factsheet, RFE/RL, July 1992, p. 8.

16. Mircea M. Dabila, *Europa* (April 6–13, 1993).

17. Adam Michnik, *The Guardian* (March 1, 1991).

18. "Surrogate Broadcasting: An Evolving Concept," RFE/RL.

19. Foreign Relations Committee Report on the Foreign Authorization Act, fiscal years 1994 and 1995.

20. "Additional views of Senator Biden" are part of the Foreign Relations Committee Report on the Foreign Relations Authorization Act, fiscal years 1994 and 1995, p. 74.

Glossary

ABC   Australian Broadcasting Company.

ABU   Asia-Pacific Broadcasting Union.

AFRTS   U.S. Armed Forces Radio and Television Service.

ARABSAT   Arab Satellite Communications Organization.

ASBU   Arab States Broadcasting Union.

ASIASAT   Asia Satellite Telecommunications Company.

BBC   British Broadcasting Company.

BBCWS   British Broadcasting Company World Service.

BSS   Broadcast Satellite Service.

CATV   Community Antenna Television—cable TV.

CBA   Commonwealth Broadcasting Association.

CBC   Canadian Broadcasting Company.

CCTV   China Central Television.

CIS   Commonwealth of Independent States.

CNN   Cable News Network. Worldwide cable news service.

COSSAT   Communication Satellite Corporation. Established in the United States in 1963.

COPUOS   Committee on Peaceful Uses of Outer Space.

CPBS   China People's Broadcasting System.

CRTC   Canada Radio-Television and Telecommunications Commission.

DBS   Direct-broadcast satellite.

DCRB   Department of Culture, Recreation, and Broadcasting. Controls Hong Kong broadcasting.

Deutsche Welle   Germany's external broadcast service.

EBS   Ethiopia Broadcast Service.

EBU   European Broadcasting Union.

EC   European Community. Body monitoring commercial broadcast practices.

ECS   European Communication Satellite.

ERT   Greek Radio Television. Oversees all broadcast operations.

ESA   European Space Agency.

EUTELSAT   European Telecommunications Satellite Organization. External service Broadcasting service aimed at other countries.

FCC   Federal Communications Commission. U.S. electronic communications regulatory body.

France Telecom   National French satellite network.

GATT   General Agreement on Tariffs and Trade.

Globo   Brazilian broadcasting conglomerate.

HDTV   High-definition television.

HONDUTEL   Honduras federal broadcasting agency.

IAHCR   International Association of Mass Communication Research.

IBA   Independent Broadcasting Authority, in the United Kingdom.
INMARSAT   International Maritime Satellite Organization.
INSAT   India satellite system.
INTELSAT   International Telecommunications Satellite Organization.
INTERSPUTNIK   Russian satellite link.
ITA   Britain's commercially oriented Independent Television Authority.
ITC   Independent Television Commission. Public agency in the United Kingdom responsible for licensing and regulating TV.
ITU   International Telecommunication Union. Operates through the United Nations.

JRT   Yugoslavian broadcast coordinating body.

LCD   Lowest common denominator.

NAB   National Association of Broadcasters, in the United States.
NASDA   National Space Development Agency of Japan.
NHK   Japan Broadcasting Corporation.
NOS   Nederlandse Omroep Stichting. Coordinates Netherlands production and air facilities.
NTSC   National Television Standards Committee of the United States.
NTV   Chile's government station.
NWIO   New World Information Order.

ORF   Osterreicher Rundfunk and Fernsehen. Austrian government agency overseeing telecommunications.
ORTF   Office of Radiodiffusion-Television. France's broadcast governance agency.
Ostankino   Russian broadcast regulatory agency.
OTS   Orbital Test Satellites.

PAL   Phase alternation by line. For TV transmission.

PANAMSAT   Pan American Satellite.
PTT   Post, Telephone, and Telegraph.

Radio Beijing   Chinese external broadcast service.
Radio Free Europe   U.S. external broadcast service.
Radio Liberty   U.S. external broadcast service.
Radio Marti   U.S. external broadcast service.
Radio Moscow   Russian external broadcast service.
RAI   Italian Radio Audition. Operates public networks.
RTC   Mexico's broadcast regulatory agency.
RTE   Radio Telefis Eireann. Ireland's broadcasting authority.
RTHK   Radio Television Hong Kong.
RTL   Radio-Television Luxembourg.
RTP   Radio-Televisoa Portuguesa.
RTVE   Spain's Radio-Television Espanola.

SSR   Societe Swiss de Radiodiffusion et Television.
SECAM   Television broadcasting standard used by several nations, primarily in Europe.
SES   European Satellite Society.

TVP   Television Poland.

UAE   United Arab Emirates.
UAR   United Arab Republic.
UNESCO   United Nations Educational, Scientific and Cultural Organization.

Voice of America   U.S. external service.

WARC   World Administrative Radio Conference. Key arm of ITU, scheduling special meetings and conferences.
WRTH   World Radio Television Handbook.

YLE   Finnish Broadcasting Company.

Further Reading

Abundo, Romeo. *Print and Broadcast Media in the South Pacific.* Singapore: AMIC, 1985.

Akwule, Raymond. *Global Tele-communications.* Stoneham, MA: Focal Press, 1992.

Alexandre, L. *The Voice of America: From Detente to the Reagan Doctrine.* Norwood, NJ: Ablex, 1988.

Alisky, Marvin. *International Handbook of Broadcasting.* New York: Greenwood Press, 1988.

———. *Latin American Media: Guidance and Censorship.* Ames: Iowa State University Press, 1981.

Akwule, Raymond. *Global Tele-communications: Technology Administration and Policies.* Boston: Butterworth-Heinemann/Focal Press, 1992.

Boyd, Douglas A. *Videocassette Recorders and the Third World.* White Plains, NY: Longman, 1988.

———. *Broadcasting in the Arab World.* Philadelphia: Temple University Press, 1982.

Briggs, Asa. *The BBC: The First Fifty Years.* Oxford: Oxford University Press, 1985.

Browne, Donald R. *Comparing Broadcast Systems: The Experiences of Six Industrialized Nations.* Ames: Iowa State University Press, 1989.

———. *International Radio Broadcasting: The Limits of the Limitless Medium.* New York: Praeger, 1982.

Burke, Richard. *Comparative Broadcasting Systems.* Chicago: Science Research Associates, 1984.

Cook, Thomas W., ed. *Communication Ethics and Global Change.* White Plains, NY: Longman Inc., 1989.

Dyson, Kenneth, and Peter Humphreys, eds. *The Politics of the Communications Revolution in Western Europe.* London: Frank Cass, 1986.

Eugster, Ernest. *Television Programming Across National Boundaries: The EBU and OIRT Experience.* Dedham, MA: Artech, 1983.

Fortner, Robert S. *International Communication: History, Conflict, and Control of the Global Metropolis.* Belmont, CA: Wadsworth Publishing, 1993.

Frederick, Howard H. *Global Communication and International Relations.* Belmont, CA: Wadsworth Publishing, 1993.

Gerbner, George, and Marsha Siefert, eds. *World Communications: A Handbook.* New York: Longman, 1984.

Head, Sidney W. *World Broadcasting Systems: A Comparative Analysis.* Belmont, CA: Wadsworth Publishing, 1985.

———. Christopher H. Sterling, and Lemuel B. Schofield. *Broadcasting in America: A Survey of Electronic Media.* 7th ed. New York: Houghton Mifflin Co., 1994.

Howell, W. J., Jr. *World Broadcasting in the Age of the Satellite.* Norwood, NJ: Ablex Publishing, 1986.

Johnston, Carla B. *International Television Co-production.* Boston: Butterworth-Heinemann/Focal Press, 1992.

————. *Winning the Global TV News Game.* Boston: Butterworth-Heinemann/Focal Press, 1996.

Kuperus, Bart, ed. *WRTH Satellite Broadcasting Guide.* New York: Billboard Books, 1994.

Matelski, Marilyn. *Vatican Radio.* Westport, CT: Praeger Publishing, 1995.

McCavitt, William, ed. *Broadcasting Around the World.* Blue Ridge Summit, PA: TAB Books, 1981.

McLean, Mick, ed. *The Information Explosion: The New Electronic Media in Japan and Europe.* Westport, CT: Greenwood Press, 1985.

McQuail, Denis, and Karen Siune. *New Media Politics—Comparative Perspectives in Western Europe.* London: Sage, 1986.

Nordenstreng, Kaarle, Enrique Gonsales Manet, and Wolfgang Kleinwachter. *New International Information and Communication Order.* Prague: International Organization of Journalists, 1988.

Rogers, Everett, and Francis Balle. *The Media Revolution in America and in Western Europe.* Norwood, NJ: Ablex, 1985.

Tydeman, John, and Ellen Kelm. *New Media in Europe.* London: McGraw-Hill, 1986.

Verna, Tony. *Global Television.* Boston: Butterworth-Heinemann/Focal Press, 1993.

Index